DATE DUE

APR 1 0 1998		
MAY 1 3 2003		

DEMCO 38-297

OPEN ROAD

A Celebration of
the American
Highway

PHIL PATTON

Simon and Schuster
New York

Copyright © 1986 by Phil Patton
All rights reserved
including the right of reproduction
in whole or in part in any form
Published by Simon and Schuster
A Division of Simon & Schuster, Inc.
Simon & Schuster Building
Rockefeller Center
1230 Avenue of the Americas
New York, New York 10020
SIMON AND SCHUSTER and colophon are registered trademarks of Simon
& Schuster, Inc.
Designed by Levavi & Levavi
Manufactured in the United States of America
10 9 8 7 6 5 4 3 2 1
Library of Congress Cataloging in Publication Data
Patton, Phil.
    Open road.

    Bibliography: p.
    Includes index.
    1. Roads—United States—History.   2. Roads—Social
aspects—United States.   I. Title.
HE355.P38   1986   388.1'0973   86-1832
ISBN: 0-671-53021-6

Contents

Three: The All-American Roadside

Four: Gone to Look for America

OPEN ROAD

Preface

Some years ago, John Brinckerhoff Jackson, the father of American landscape study, argued that we needed a science of roads—"odology" he would call it, after "hodos," the Greek word for road.

This book is an essay in search of that imaginary science. It looks at American roads as artifacts and as shapers of the landscape. It is about their architecture and engineering, but also about their politics and economics. But above all, it is about the special status of the road in the metaphorical landscape of the American mind. Some of this mental landscape is conscious; a good deal more is unconscious, romantic and mythic. There is nothing more American than being On the Road.

Part One tells the story of the creation—the invention—of the American highway. Part Two explores its design, the internal logic of its shapes, the physical devices by which it maintains the ideal of the flow. Part Three looks at the landscape the highway created along its sides. And Part Four traces the imaginative landscape of the highway, the dream and mythology of the open road.

*"The Road West." New Mexico, 1938, by Dorothea
Lange. En route to the Pacific Coast on U.S. 80.
(THE OAKLAND MUSEUM)*

Introduction:
"Oh Public Road!"

The road runs straight across the open plains, toward a bright horizon where mountains or mesas rise, squeezing the white stitching of its centerline together in a caricature of perspective.

The sun is setting but heat still shimmers from the black asphalt, warping the horizon, and the sky which nearly fills the windshield is already waxing rosy beneath a higher blue. A telephone pole, crossed with four or five arms, stands out against the sky like a musical staff marking the bare sound of rushing wheels.

Soon the evening breezes will nudge tumbleweeds across the pavement and coyotes will begin to slink up to the shoulders of the road, hiding from the yellow cones of the first headlights in the ditch where a rusting car hulk, a beer can, and a forgotten hat lie together.

We all know this road. It is the quintessential American road, the simplest, straightest road, a road in the center of the country, heading west, a road in between places. Dorothea Lange and Robert Frank photographed it. Steinbeck and Kerouac wrote of it. Country music sent its maudlin, lonely, seeking, escaping heroes down it.

A hundred movies have cast it, in rear projection, as backdrop. It is permanently imprinted on the national consciousness. It looks like Route 66, but it could also be Route 80 or Route 30. It could be in New Mexico or Colorado, Idaho or Wyoming.

Migrants, looking for a better life, took the road. So did adventurers, vacationers, and killers: Clyde and Bonnie Barrow, Charlie Starkweather.

Those who drive this road are often alone and often between identities. They are abandoning old lives and looking for new ones, but are most themselves in the interval. Some drive to remember, others to forget.

This road, like all American roads, offers a literally concrete expression of the central American drives. If, as Oscar Wilde quipped, America's youth is its oldest tradition, change is its most unchanging premise, movement is its most firmly fixed pattern, impermanence its most permanent condition, and the receding horizon its most steadfast goal. And America's restlessness is its most mercurial trope.

The mute perspectives and pavements of the highway objectify elements of the American mind. We speak of the American way of life, not the American form of life. The road echoes the fundamental observations of Crèvecoeur, whose "new man" was constantly restoring his novelty in movement, of Tocqueville, who found Americans at ease only in restlessness, of Emerson's "there is no truth but in transit," of William James's stream of consciousness. The noise of the road lives in the rhythms of its prose and song, the throbbing verbals of Faulkner and Agee, the bubbling expressionism of Kerouac.

"Oh public road," intoned Whitman in his *Song of the Open Road*, "you express me better than I can express myself."

Whitman's road was the track of an abstraction—"I believe you are not all that is here,/I believe that much unseen is also here."

Even before the automobile, American roads were as much idealized roads as physical ones. With its advent they became, from dirt roads to the Interstates, basic social and cultural expressions, artifacts of our national obsession with mobility and change, with the horizon, with the frontier.

Roads are the creation of collective design. They reflect unconscious as well as conscious patterns of politics, economics, and culture. American roads reflect a fundamental political and philosophical conflict between the systematic and the spontaneous, the national and the local, between roads from above and roads from below.

American roads have shown the face of the country, its inventiveness, its fantasies, and its follies. Along their roads, Americans have both sought and fled identity, carrying out their pursuit of happiness. What Steinbeck wrote of the Okies had echoes in the whole national sensibility: "The highway was their home and movement their means of expression."

Highways have made tangible the conviction that the truth about America, its heart and soul, collective and individual, is always to be found somewhere just over the horizon, somewhere around the next bend.

Roads are a realm of signs, a set of clues to the constantly receding mystery of nationality. The culture they have fostered has become a physical model of the fact that the "promise" and "potential" of America have congealed into a permanent system offering one set of promises after another, with the easy recession and happy forgetfulness of a moving perspective.

The automobile and its highways froze the values of the frontier by making movement a permanent state of mind, turning migration into circulation. The wave of migration, having carried to the West Coast, rebounded with the closing of the frontier, the completion of the expansion, into an echoing succession of movements. These movements became a social and a personal hunger. The average American spends more on transportation than he does on food.

Transportation has always had a dominant role in the American economy. We built canals, railroads, and finally highways in advance of, not in response to, the areas they served. Thus, more than elsewhere, each new system of transportation shaped the economic and social landscape rather than being shaped by it. And they required special forms of social and economic organization: they demanded a government involvement that was otherwise foreign to our political culture.

Almost from its beginning, the automobile was seen as a pleasure machine, meant to increase social opportunities, to provide adventure and access to nature and, soon, to a whole range of amusements created exclusively by the highway culture.

Road travel has always been entertainment tinged with nostalgia. The autocamping fad that swept the country in the early twenties— making, at one time, Henry Ford, Thomas Edison, Harvey Firestone, John Burroughs, and President Harding companions on a weekend Model-T jaunt—was succeeded by the creation of the entertainment strips, the great midways of make-up adventures accessible by motorcar.

The waterway and the railroad, the two previous dominant stages of American transportation, each provided their cultures: the Jolly Flatboatmen and Casey Jones, *Life on the Mississippi* and *The Bride Comes to Yellow Sky*, the canny pilot and the heroic engineer, the riverboat and the locomotive, the riverside village and the "Hell on Wheels" town. To these, the motor road added its own culture. Along their sides, American roads introduced a whole mad, sad, crazy, funny carnival of culture: temporary architecture, a car-toon architecture of fantasy, abbreviation and pun, slang and contraction, on an art combining surrealism, pop, and cuteness.

The strips—great assemblages of American miscellany, long exclamatory catalogs, like Whitman's, facts posing as poetry—took on a life of their own. They were asphalt jungles with their own ecology, communities of commerce and information with their own law and rules of survival.

The modern superhighways tried to sweep away this roadside carnival. They celebrated the streamlining of our culture, the smooth flow of the "mainstream," bypassing the tributaries of old local eccentricities. American roads have always been more about the past and future than about the present. The Interstates and freeways were the roads of the future; their construction turned all other roads into byways of the past, objects of nostalgia.

Modern highways have made us look at landscape in a new way: cinematically, from the carefully tracked panoramic shots of the highway. At 65 miles an hour, experts say, the driver sees five times as much sky as at 45. Roads are drunk with the principles of perspective. They ignite the vast spaces of the West, offering a vantage point from which to mark and appreciate the recession of desert, plain and prairie. Approaching the city, they convert its towers into sculpture provided for the exhilaration of the driver.

In design, they offer a virtual parody of form following function: their shapes are drawn by the physics of motion. The curves of their structure describe the curves of their use. They are classic modern architecture, dedicated to the beauties of steel and concrete.

Roads sometimes seem in danger of vanishing into familiarity, of disappearing into their omnipresence. We have long focused on the automobile to the exclusion of the highways it travels. The automobile, after all, is to most of us an extension of our self-image, of individual style, of clothing—the speed with which the auto industry became a fashion industry, the black puritan uniformity of the Model T gave way to the annual model change and market spectra of General Motors shows this. All the cheap and obvious psychologizing about the automobile as castle on wheels, mobile mating den, symbol of power and status ignores the fact that the automobile was determined by where it goes. It is like looking at the television set without the shows, the phonograph without the records, the computer without the programs, the hardware without the software.

The relationship of the car to the road can be understood by turning around a comparison tossed about in the world of electronics. The mainframe computer, this analogy holds, is like the railroad: it is centralized, controlled by large companies that own both hardware and software. The individual can use it only by acceding to a fixed schedule—timesharing—and is often subjected to being "folded, spindled, and mutilated" by its operation.

The micro-computer, by contrast, is like the automobile. It allows the individual to choose his own routes and schedules; it encourages private initiatives and the creation of new forms of accessible, inexpensive, or free "public domain" software. It links individual uses to industry standards.

The success of the micro-computer depends on programs: the success of the personal automobile depended on highways and their adjacent commerce.

We tend to speak of "the impact of the railroad"—denoting track, locomotive, Pullman car in one word—but of "the impact of the automobile," including the road only implicitly. But it is the highway that has been the direct agent of changes in our physical, economic, and social landscape.

The mythology of the automobile began from the premise that it was a railroad without rails, that it could go anywhere. Part of the reason for the success of the Model T was that it could handle the roughest roads, that it was the car that came closest to the ideal of going anywhere.

Assume, as Henry Ford did of the car, that soon "no one will be unable to own one," and it is a short step to "no one will be able *not* to own one." Assume that the automobile can go anywhere and soon it *must* go anywhere: you have the impetus for the great roadbuilding schemes. We have built roads even into Yellowstone and Big Bend because we would no more imagine people venturing into the parks without cars than without shoes. Roads mark the extent of our civilization: when Congress passed the Wilderness Act in the mid-sixties, wilderness was legally defined as any contiguous area *at least five miles from the nearest road.*

The appeal of the automobile depended on a vision of an infinity of roads awaiting the driver, of magic possibilities unfolding without end. But we thought more about the car than about the roads, as if the roads would spring up naturally. It was an idea as naive as Whitman's: for we create the roads, as we do the automobiles— finned, chromed, downsized, or elongated—as a society. As design, architecture, politics, economics, engineering, highways are fundamental artifacts of our culture.

Roads are social models at least as much as buildings or parks, a sketch of how we deal with human freedom and interaction, human ability and inability. Roads are designed around the abilities and limitations of the individual driver, the generalized license holder, the democratic atom of the mainstream. The fast lane has become a metaphor for a lifestyle; merging lanes represent an etiquette in miniature, a ship-in-the-bottle version of the social contract.

The whole phylogeny of federal-state relations is recapitulated in the ontogeny of every Interstate highway project. And the whole legacy of American ambivalence toward cities is present in the patterns of freeway construction in those cities.

Our most ambitious road programs expressed a conviction that building highways can do nothing less than expand the economy, reshape society, lift up the poor and lowly, draw regions together, blend city with country, and surpass distance by melding it with time.

The Interstate program, the most expensive and elaborate public works program of all time, offers a vision of social and economic engineering. It was planned to be at once a Keynesian economic driver and a geographic equalizer, an instrument for present prosperity and the armature of a vision of the future. It was at once the last program of the New Deal and the first space program.

The Interstates and other superhighways fueled the growth of the Sunbelt. Their rise paralleled and abetted the decline of the railroad-based industries of the Frostbelt and ripped apart the fabric of the old cities. It was in 1958, as the Interstate program began to take effect, that the service sector overtook manufacturing as the largest component of the economy.

The Interstate program served practically every large interest in the country, from the automobile industry, with its backsuppliers in steel and coal, to the petroleum and chemical industry, and all levels of government bureaucracy. It brought government home to the average citizen with the physical tangibility unprecedented except by service in the military: "Your tax dollars at work."

Once, for instance, the immigrant's first encounter with his new government was through admission centers established in disused forts and arsenals: Castle Clinton on the Battery or Ellis Island. But when the Cuban refugees of the Mariel boatlift arrived in Miami, they were quartered and "processed" in camps constructed on the most available federal property—the right of way beneath elevated portions of Interstate 95. The space was an appropriate one: the highway itself, and the Cadillacs and Chevrolets, the Mercedes and Toyotas, the Jaguars and Broncos whizzing by over their heads were promises to the new arrivals of the American possibilities of prosperity and mobility—and illustrations of their limits.

At the height of the Roman Empire, a map of hammered gold hung in the palace of Caesar, embossed with the lines of the Roman road system, with the principal cities of the Empire they connected appropriately represented by rubies, emeralds, and other precious stones. This map expressed a system that belonged to Caesar, built by his legions as the military and communication tendons holding his empire together.

Another map of the Roman highways stood in the Forum, carved in marble, with the same pattern conveying a different message: that the roads were a great public achievement, like the temples in the Forum, a masterpiece of civilization triumphing, like the alphabet itself, over barbarism.

We have no such central map of our highway system. Where would it be? Our Caesars Palace, having abandoned the possessive apostrophe as if in token of collective and anonymous ownership, is only the most famous building on our most famous Strip, a monument to the individual's good or bad luck. Our capital is a warren of bureaucracy; the highway section of the Department of Transportation sports a map of the Interstates, but the Interstates, like the rest of our roads, are for the most part designed locally, to national standards that are at best abstract.

In the Empire, of course, all roads led to Rome. In America, all roads lead to other roads. None of our roads have satisfactory ends: Interstate 80 does not debouch grandly onto the beach at Marin, or divide around a grand marker in Times Square.

Our roads were not laid out in Washington, but in the various state capitals and even county seats. Despite the myth that our roads followed wagon trails, which followed Indian paths, which followed the tracks of animals, our roads in fact were laid out by surveyors. Our maps reside in glove compartments, once placed there compliments of oil companies. Now they come to us as premiums from automobile associations, or are picked up at convenience stores, beside the snack foods. Before long, they will arrive in more disembodied form, over computer equipment linked to satellites or earthly towers.

If the Roman roads were a hub and its spokes, ours are a grid, their deviations from absolute rigidity a mere accommodation of geographical accident. Our roads are considered to be of equal importance at any given place: they assert democracy, the power of the people, the service of the public—and the statistical profile of the average driver, average consumer. The limo and the jalopy are equal in their sight.

The Roman road system was at once the property of the emperor and the empire, which were inseparable. Like earlier systems, it was chiefly a way to get troops and official messengers about, and most subsequent road systems would be the same. The highway was the king's highway.

The word *road* implied something more natural than highway. *Highway* held intonations of royalty and aristocracy; a high road implied a low road. The term *highway* was revived with the arrival of the automobile, as if to signal that every driver was king of the road. The American highway, by contrast to the imperial roads of the past, was a common carrier.

Until the telegraph, roads were a means of communication as well as transportation. They carried both courier and legion, brought news from the far reaches of the empire, and delivered the message of military power and control. There were differences in speed, more marked as time went on—the pony express could move much faster than the cavalry—but the means of delivery was the same. In this country, the first improved roads were dedicated to the post. But by the mid-twentieth century they had become places *to be*. Real estate men began to speak of highways that had become "saturated," their sides and interchanges "built to capacity." Roads were no longer even primarily means of transportation: they were patterns for location.

Above all, our roads are public spaces. "Public road"—Whitman's inelegant, bureaucratic term was current from the 1880s to the 1960s. Public roads, before this century, were mostly inferior substitutes for private ones: the most sophisticated roads of the nineteenth century were private turnpikes, and the railroads were owned by giant corporations. Only the automobile made all roads public.

Like television, American highways are a national network, a mass medium. Driving, in fact, has been studied by psychologists as a process not unlike watching television. As we bemoan the effects of television, we also lament the effects of modern highways—with almost the same intensity with which we devote ourselves to them. They have had monstrous side effects. They have often rolled, like some gigantic version of the machines that build them, through cities, splitting communities off into ghettos, displacing people, and crushing the intimacies of old cities with a scale taken from dreams of the wide-open spaces. They have become a symbol of the American distrust and dislike of the city.

And yet American roads are a reservation of individuality and privacy. Americans have gone on the road to find America and find themselves, following in the naive Whitmanesque mode. The song of the open road is also the song of the self, a song of elusive identity, of blue skies, blue roads, and the blues.

While promising to bring us closer, highways in fact cater to our sense of separateness. The quintessential American, like the basic country singer, who is constantly singing about moving on down the road, taking the lost highway, is a connoisseur of special loneliness. The man out to see the USA in his Chevrolet, with or without family, cultivates a sense of alienation. His vacations are a continual reenactment of discovery and exploration, and for discovery, we must refresh our sense of being aliens, nurture our otherness.

Ultimately, any set of roads embodies a philosophy, a "way." How many American dreamers, in search of the national spirit and ultimately themselves, have taken to the road with the same bluster as Whitman, to find Truth and the Way and Ourselves and Themselves? Writers and tourists both have taken to the road, dedicated to the proposition that on the road the traveler must inevitably, eventually, meet himself. "To know the universe itself as road," Whitman writes: a tall order, a romantic ambition, an American dream.

The road's cult of the individual approaches the pathological. American society has found the death by accident of some 50,000 people a year on the roads—with a cumulative total that long ago passed the common comparison, "more than in all wars the United States has fought"—an acceptable price to pay for preserving the individuality of the driver on the common ground of the highway. And we struggle along blindly in the face of immutable but incomprehensible laws by which each new highway seems to fill to capacity as soon as we can complete it. And yet we continue to dream of new highways, safe and free flowing, open roads.

For in America, highways are much more than a means of transportation. They come as close as anything we have to a central national space. They are a national promenade, "America's Main Street," and a medium in which grows the carnival of individual life and enterprise. While the Romans had their road map in the Forum, for Americans roads are themselves the Forum.

WESTON, *Brett.* New Highway North
of Santa Fe. *(1938) Gelatin silver print,*
7⁹/₁₆ × 9¹/₂" (19.3 × 24.0 cm).
COLLECTION, THE MUSEUM OF MODERN
ART, NEW YORK. GIFT OF DAVID H.
McALPIN

ONE

From the Ground Up

The first rural mile of concrete highway, completed in July of 1909, was part of Detroit's famous Woodward Avenue. (*MICHIGAN DEPARTMENT OF TRANSPORTATION, PHOTO SECTION*)

1: *The Road into
the Wilderness*

*"Let us then bind the republic together with a perfect system of
roads and canals. Let us conquer space . . ."*

—*John C. Calhoun 1816*

In 1803, with an eye to securing communications with the Mississippi Territory, President Thomas Jefferson ordered the improvement of the Natchez Trace. It was one of the first, and one of the few, roads built—if the slashing of underbrush and hacking of stumps by a few companies of soldiers can justify the word—by the young federal government.

Jefferson was very specific. He ordered that

> between the Grindstone Ford and the Chickasaw towns, where 18
> to 25 miles have been lost . . . to avoid swampy lands, a resurvey be
> made to see whether these might not be made passable . . . all streams
> under 40 f. not fordable at their common winter tide shall be bridged;

> & over all streams not bridged, a tree shall be laid acros, if their
> breadth does not exceed the breadth of a common tree. Where they
> exceed that must a boat be kept?

This attention to detail suggests just how much Jefferson invested
in the value of roads and waterways to secure communication with
the new territories—a vision that grew stronger after the Louisiana
Purchase of 1803.

Washington and Hamilton, among others, had been conscious
of the need to bind a country as large as the United States together
with means of communication and transportation, not just for economic reasons but for political ones: communication was essential
in a democracy. Jefferson, despite all his differences with the Federalists, agreed, and after him Clay and Calhoun boosted "internal
improvements" for the sake of the West and South and a new
nationalism. "The mail and the press are the nerves of the body
politic," Calhoun argued in 1817, and they depended on internal
improvements.

Space was abundant in the new country; time was at a premium.
Reduce the amount of time required for communication and transportation and the country's vastness lost its disadvantages and began
to realize its benefits. In such a situation, roads became more than
simply a means for moving goods and traffic. They became an
almost mystic factor of American political life.

The stretches of the Trace that remain today are among the most
striking of the few surviving pieces of early American roads: the
remnants of the Philadelphia-Lancaster Pike of 1792, once the finest
road in America; a few tollhouses, each with its "pike" barring the
traveler's way and giving the turnpike its name, and often a tower
room for the tollkeeper to scan the road; a few mileposts—sturdy
engraved stones, including the famous "Franklin stones" set up
along the old Post Road (now U.S. 1) between Philadelphia and
Boston at the order of Benjamin Franklin; a few magnificent stone
bridges, like the arched stone bridge over the Monocacy River near
Frederick, Maryland or the Casselman Bridge near Grantsville,
West Virginia.

The Trace, however, seems almost primal. Its main course was a Choctaw and Chickasaw hunting and war trail. There are places where the traffic of subsequent boatmen, armies, and settlers literally drove it into the ground. There it runs through a wide trench, perhaps twenty-five feet deep, romantically arched over with trees.

The brief golden age of the Trace was created by the traffic of "Kaintucks," as settlers of the new territories west of the Appalachian Range were called, regardless of their actual state of residence, frontier farmers who floated their goods down the Tennessee and Ohio and Mississippi to Natchez and then had to walk or ride home. Often they employed the "ride and tie" system, two men sharing a mule or horse. One man would ride ahead a distance, tie up the mule, and walk. His fellow would reach the mule on foot, mount it, and catch up.

It was a dangerous trip. The Trace existed only at the sufferance of the Choctaws and Chickasaws and outlaws who hid out in natural dens like the Great Bridge or in sinkholes, menacing travelers. The few inns, called "stands," were little more than cabins where gamblers and cheap whiskey were to be found in abundance. The biweekly "Great Mail" to Natchez—Jefferson's immediate justification for improving the Trace—not infrequently disappeared. The rider who carried it was blandly noted in postal records as "presumed lost."

As a free government road the Trace was exceptional. As an "improved" road it was even more so. Most roads were like trails, "given" by the landscape, taking their course from watersheds, ridges, and mountain gaps. The idea of *building* roads was still so strange in a nation of trails and traces that Secretary of the Treasury Albert Gallatin described them in his "Report on Roads and Canals" of 1808 as "artificial" roads. The word *road* took on a highly abstract, almost idealized meaning: it referred to any line of transport, on land or on water.

Gallatin's landmark plan was the Jefferson administration's vision for a major federal system of internal improvements. Jefferson's administration began with the aim of undoing Federalist interventions. Jefferson and Gallatin had liquidated the national debt, and the internal improvements program marked the single positive proposal of this first Republican administration, the only affirmative role for government. There was an irony to the fact that an administration dedicated to limiting the power of the federal government should propose a system of internal improvements far more dramatic than anything suggested by its Federalist opponents. The plan marked the beginning of a curious association between the conservative side of American politics and the construction of roads. Even so, Jefferson believed a constitutional amendment was necessary to give the federal government specific power to carry out the construction he proposed.

Jefferson's aim was to foster communications precisely so that the country would not become too centralized, too dependent on the eastern cities he despised. To the centripetal power of maritime trade and manufacture, he proposed a centrifugal power of westward expansion, easily acquired land, and good transportation.

Gallatin's ten-year plan included turnpikes, canals, and waterways. A great turnpike was to run from Maine to Georgia. Four turnpikes were to cross the Appalachians to the western rivers, and canals were to link the rivers into waterways. The program was to cost a relatively modest $2 million a year—less than fifteen percent of the budget, and an amount the federal budget surplus of the day could easily cover. Sale of stock in turnpikes or canal companies could foster further improvements, and proceeds from the sale of the western lands could be used to extend roads there, as had already been stipulated in Ohio.

Gallatin viewed his plan as a response to the particular economic and geographic situation of America. He had already had occasion to see the critical importance of transportation for binding the Union. Gallatin was from western Pennsylvania, where, in 1794, the federal government had faced the first serious challenge to its authority: the Whiskey Rebellion. Bad roads, it could be said, had caused the rebellion, which was set off by the imposition of a whiskey tax. To Hamilton and the Federalists, the excise tax on whiskey that sparked the resistance seemed a levy on luxury, if not sin. But to the western farmer, there was no way to get his corn to eastern markets except by turning it into whiskey. Whiskey was lighter, it didn't spoil, and it was easier to transport.

The federal government had to build the roads that would prevent such future disruptions, Gallatin argued. In Europe, private fortunes were well enough established to build canals, roads, and other improvements on their own; not so in a young country like the United States.

It was significant that Gallatin spoke of highways almost strictly in what would today be called cost analysis terms. He noted that improved highways would make communication easier, but most of the benefits he saw were couched in simple economic terms. This was a long way from the simple expedients of absolute power, projected over the straight line, that had inspired the Roman Caesars or French kings to build highways.

But Gallatin's plans were rejected, largely due to rivalry among the states and to the onset of the War of 1812. Improvements to the Trace, part of a road to Detroit, and the Cumberland (or National) Road, begun in 1807 before the presentation of the plan, were the only parts of it realized. The National Road ran from the head of navigation of the Potomac, near Cumberland, Maryland, to Wheeling, on the Ohio River. (The route is now that of U.S. 40.) It was intended to help open up the Northwest Territories to trade and development. Financing came from sales of federal land in Ohio. Eventually extended into Indiana, the National Road was to become a vital thoroughfare, but it was unique: *the* National Road.

The war that thwarted the proposals, however, also presented a strong argument in their favor. During the War of 1812, the government had great difficulty moving troops to the northwest and south. Andrew Jackson had to pay a private ferryman an outrageous sum to carry his army across a river on the unimproved Trace. And the need for the Cumberland Road to the west was so evident that it was completed on a crash program within two years after the end of the war. Sixty feet wide, of crushed stone fitted in graded sizes, it was the best road in the country.

For the most part, road building in nineteenth century America was left to private corporations—whose financial organizations, incidentally, helped establish the model for railroads and other later corporations—chartered and sometimes partially financed by state governments.

To Jefferson, the importance of building new roads and improving those like the Trace lay not only in the immediate pressures of geopolitics, such as the need for a line of approach to the critical port of New Orleans. His vision of roads into the territories was closely tied to his whole political philosophy. Americans were to pursue happiness down roads that led to inexpensive land in the west. There they would become yeoman farmers, husbandmen, ploughmen, for land was not only the only natural source of wealth, but in contrast to the corruptions of the city and its manufactures, the only morally healthy locale.

The more its people owned land, Jefferson believed, the stronger the republic would be. Land was the basis of Jefferson's democracy. He believed that only landowners should vote, but he also believed that government should give land to any landless citizen.

Land ownership in the United States was a much stronger institution than in Europe—a fact that was to shape the development of American roads right through the Interstates, when the claims of ownership were defined as requiring compensation for relocation, loss of access, noise or visual impairment, and other consequences of construction. Jefferson, working to abolish such old-world restrictions on land ownership as primogeniture and entail, the survivals of feudalism, helped create that American brand of ownership.

Jefferson's view of roadbuilding—indeed of all improvements—was a larger version of his ideal of the farmer, the "husbandman" or "ploughman," to use his vocabulary. The notion of the husbandman as one whose role is to assist the natural fecundity of the land (his "wife") is a traditional one. He could do this better with, say, a better plough; Jefferson invented one with a non-stick mouldboard.

A road, then, was like a plough: it aided in the cultivation of the new territories. The natural roads and waterways of America could be assisted by cultivation: improving the "natural" course of the

Indian and Kaintucks' trail to create the Natchez Trace, linking waterways with canals where practicable, building freight roads over the mountains. Road building would be the successive stage to the reconnaissance of new territories by Meriwether Lewis and William Clark. It would deploy the first patterns of the furrow across the land.

Measuring, dividing, categorizing were favorite tools of the Enlightenment—think of Buffon's classification of species and the *Encyclopédie*, the categorized compendium of number and definition of which Jefferson's *Notes on Virginia* is a miniature version. Jefferson, wrote Richard Hofstadter, "had an almost compulsive love of counting, observing, measuring."

Jefferson was largely responsible for the grid organization of American land, which provided the basic framework for the development of all American roads.

The grid was the apotheosis of the surveyor, of the power and ethos of the theodolite that shaped our landscape. Washington was a surveyor, Jefferson's father was a surveyor, and Jefferson knew the science—it speaks clearly in his instructions for the construction of the Trace.

Most American surveyors were self-taught. They relied on a critical but largely unsung invention: Gunter's chain, the creation of Edmund Gunter, an English surveyor. This linked steel chain was 22 yards long; an acre was defined as ten square chains. Gunter's chain combined the square unit with the decimal system—which Jefferson, always the measurer, was responsible for introducing to the U.S. currency.

The grid shaped almost all of the American landscape outside the thirteen original colonies. It was an abstract device, an order imposed on the landscape. In France in the seventeenth and eighteenth centuries, a triangular grid was attempted, linking parishes and political units, and roads were defined between that grid and the straight radii, centering on Paris, of Louis XIV's highways. But this system never possessed the power that the American grid system did over its landscape.

The American grid was institutionalized in the Land Ordinance of 1785, of which Jefferson, although his ideas were modified, was a key author, and the Land Act of 1796, which laid out procedures for the sale of government lands in the territories. The Land Act divided lands into townships six miles square, made up of 36 sections, each of 640 acres. With the townships and countries it mandated, the American grid provided a physical structure for generating new political structures.

The law also required that the section lines be reserved for roads. In some states, until quite recently, law forbade the construction of any highway that did not follow the grid system—a legacy of farmers fearful of having fields sliced by diagonals. Even today, flying over the Midwest, the resulting grid of roads is quite visible, but it took years for all these roads to be built. Meanwhile, they existed as plans, as potentials, as future roads. They helped us get used to the idea of road systems whose pieces are filled in one by one, but which exist, in plan and potential, long before their physical pavement—the abstract road.

It was also significant that the grid lacked a privileged center. Even the foci of political life, the county seat and the county courthouse, were simply set on sections and blocks like any other, chosen for mere convenience. Roads in the grid necessarily covered it like a net, not in a radial or even a naturally determined pattern.

Jefferson saw the grid system as a device to help achieve his goal of land ownership by as many people as possible—part of his ideal nation of yeoman farmers, whose basic social unit was the rural county, not the evil city. In fact, the grid system as adopted tended to support land speculation, and the minimum purchase sizes forced many small settlers to buy from speculators. But the government policy of land sales, of which the grid was the organizing principle, strengthened an already powerful colonial tradition of law and economics that gave private ownership of land far wider powers than were dreamed of in English or other European systems. The consequences are clearly visible in the difficulties faced by city planners, forced to resort to such relatively blunt tools as tax incentives and zoning to limit maximum exploitation of each and every private city block.

This broad definition of the rights of land ownership provided problems for later roadbuilders. Among the property rights of the landowner was that of access to adjacent roads, while the creation of limited access highways necessarily involved the use of eminent domain. This was to have major consequences for the development of superhighways, when the government was forced to purchase outright many businesses and tracts bypassed by Interstates or other limited access roads. The old motels and diners along Route 66, for instance, bypassed by the Interstate, are decaying on expanded right of way and isolated bits of federal land.

The establishment of the grid had other, more subtle but persistent consequences. Beginning in 1916, with the first federal highway aid, the topic of a national network of highways arose; and it was the grid, however reshaped by the competing topologies of politics, economics, and geography, that served as a basis for all the versions of such networks, down to the final Interstate map.

Well before the land ordinances, American city planning, beginning with William Penn's Philadelphia, adopted the grid, mostly with the aim of retaining green, open land in the city, by contrast with dank, dark London. The most striking image of the gridded city is the famous engraving made of Savannah, Georgia, as laid out by James Oglethorpe. Like planting beds in a classical garden, the blocks of Savannah, some reserved for parks, seem to have been stamped out by some giant press among the tall pines that rise with the same linear geometry as the streets.

Grid in the city, grid over the country: these patterns dictated a necessary connection between urban street and rural highway. The grid, as the organizing principle for most cities, also tied them directly to the national highway grid.

Jean-Paul Sartre once observed that while European streets typically end in some enclosed space, the straight streets of the American city seem to run on forever, opening the very heart of the city up to the horizon and denying it the separation and special status of a Paris, Rome, or London. In those centers, the roads of the country converge; but the American urban grid disperses the city into the country.

American streets, Sartre argued, "are not sober little walks closed in between houses, but national highways. The moment you set foot on one of them, you understand that it has to go on to Boston or Chicago."

In 1948 Robert Frank photographed a major Manhattan cross street, probably 14th or 42nd Street. The center of the picture is the white line down the middle of the road, disappearing into the distance in a virtuoso demonstration of perspective. Almost all the picture is filled with pavement; only at the top edges, in perhaps a tenth of its surface, does the sidewalk and its pedestrian life, shops, signs appear, the bustle reduced as if to a distant noise. The photograph is an urban parody of the famous shots by Dorothea Lange and Frank himself of straight, receding highways disappearing across the western plains, but one that establishes something in common between the two kinds of road. It suggests the identity of American urban geometry with rural geometry, the connections of all the grids.

No wonder that the superhighway builders would venture into the city as if the roads they built across Kansas farmland or New Mexico desert could simply be transplanted to the city, perhaps raised on piles, if necessary, to allow their builders to ignore the local landscape the same way they could ignore the landscape of the wide open spaces. Both landscapes were organized on grids, but they were different grids, and the result was a debilitating conflict of scales.

Even in Los Angeles, the first city of freeways, the superhighways were organized into a loose grid. But it was a grid compatible with the street grid. The classic view of Los Angeles at night confirms the fact: seen from the hills around it, the city spreads out in a great flat reticule of lights. By contrast with eastern cities in which freeways were imposed over the grid as spokes and rims, the L.A. freeways themselves form a grid integrated into the small-scale grid of the blocks.

Jefferson's vision of internal improvements was as much in conflict with his own notions of the limits of federal power as was his purchase of the Louisiana Territory that made such improvements necessary. He could live with the contradictions; it was left to others to point them out and establish them as legal and political precedent.

The first three decades of the nineteenth century saw a succession of Presidential vetoes denying that the federal government had the Constitutional right to build and maintain roads. Almost as soon as it was completed the National Road had begun to deteriorate. When, in 1822, Congress finally passed a bill levying tolls along the road to provide for its upkeep, President Madison vetoed it. The federal government, he argued, had no right to infringe on the sovereignty of a state by levying tolls or exercising other jurisdiction on a highway within its borders without its consent.

The pattern was fixed for good in Andrew Jackson's Maysville Road veto. That Jackson should have been the president to deliver it summed up the contradictory impulses at work in the American political system—and the American character. Jackson—the champion of the West, the general who had proved the importance of the Natchez Trace by marching his troops down it to New Orleans during the War of 1812, and who later built military roads himself en route to Florida, the strong supporter of internal improvements—nevertheless found it necessary to veto the bill on the grounds that the road involved was totally within the borders of one state. There were tactical political issues involved—Martin Van Buren urged the veto as a slap in the face of Henry Clay and other administration opponents—but also a key political principle.

The bill in question authorized the purchase by the federal government of part of the stock of the turnpike company chartered by the Kentucky legislature to build a turnpike from the National Road to Maysville, Kentucky, on the Ohio River. Federal involvement, supporters argued, was justified because the route would carry mail.

While Calhoun and others had attempted to justify such improvements under the "general welfare" clause of the Constitution, Jackson insisted that a Constitutional amendment for the purpose was needed. In this he echoed Jefferson.

Such an amendment was never made; astonishingly enough, the principle of the Maysville Road veto continues in force today: *one of the most striking features of the American road system is that federal government did not officially build it.* Its money, delivered with numerous strings attached concerning location, standards, procedure, and policy, paid for the roads; but the states, counties, and cities built and own them. It is an odd, but exemplary, feature of our political culture and the culture of our roads.

In 1809, Meriwether Lewis, the man Jefferson had delegated to blaze the first trails into the distant abstraction of the grid, and then had appointed governor of the new Louisiana Territory, was called back to Washington to face a Congressional committee. On the way, he died mysteriously, of gunshot wounds, in a seedy inn along the Trace. Jefferson had a monument erected to him there: a Doric column, its shaft broken off at midpoint in token of a life cut short.

It could also have been a memorial to the end of Jefferson's vision of bold tracks into the wilderness—expressed as much in the Trace as in the Lewis and Clark expedition—delivering settlers to neat individual plots in the gridded land.

The stalemate over internal improvements kept American roads poor. Dickens, in his *American Notes*, based on a trip taken in 1842, lamented the tortures of the coach rider subjected to "corduroy roads" of logs laid side by side.

After Gallatin, no unified plan for a national system of highways was seriously proposed until after World War I, and not until the mid-twenties did the roads built with federal aid receive systematic designation.

The land that Jefferson thought would be sufficient for centuries was quickly taken up. Roads spread through it with such dispatch that when Congress passed the Wilderness Act in 1964, it officially defined wilderness as any "*roadless*" area of at least 5,000 contiguous acres—a piece of land no larger than the ambitious frontiersman might have aspired to.

Soon, the speculators and railroad men gained control of the West. Jefferson, who had been fascinated by the steam engine, especially its possibilities for locomotion, did not live to see them realized. But within a few years of his death, the steamboat had made the Natchez Trace obsolete, and it was not much longer before railroads killed the private turnpikes. Americans quickly became used to the replacement of one system of transportation with another, with a constant building and rebuilding.

Road surveying and clearing by the federal government was reduced mainly to military roads in the territories, where issues of states' rights did not surface. And the same military survey expeditions that laid out routes for the wagons of settlers heading for California or Oregon were looking for potential routes for the transcontinental railroad everyone knew must eventually be built. The course of Route 66, for instance, from Fort Smith, Arkansas, to Sante Fe, New Mexico, was blazed in 1849 by Captain Randolph Marcy and James Simpson, "in direct reference to the future location of a national road"; and the rest of the way by Lieutenant Edward Fitzgerald Beale, whose 1857 expedition from Sante Fe to the Colorado River was accomplished with the assistance of some seventy Bactrian and dromedary camels. Beale's boss, Secretary of War Jefferson Davis, found the results encouraging; they pointed to a potential southern route for the great railroad, and Beale declared that the route would "inevitably become the great emigrant road to California."

But the lands the emigrants were looking for, the lands of the great grid, went mostly to speculators and—by the millions of acres—to the railroad companies. And, like most people after 1830 or so, the military pathfinders used the word *road* as shorthand for *railroad*.

An improved section of the Lincoln Highway near
Green River, Wyoming. The transcontinental
railroad, duplicated in pavement. (BROWN
BROTHERS)

2: Boosters and Nostalgists

When Henry Joy, president of the Packard Motor Company and one of the boosters of the Lincoln Highway, tried to motor cross country early in this century, he found that the roads ended completely somewhere in Nebraska. A man in Omaha gave him instructions: drive west from town until you reach a fence. Open the fence, drive through and close it. Do this several times. Joy followed the instructions, he recalled later, until finally the fences ended and there was "nothing but two ruts across the prairie."

This was the state of American roads that, in 1912, inspired Joy to team up with Carl Graham Fisher, owner of the Prest-to-lite carbide gas auto headlight company and founder of the Indianapolis Automobile Speedway in a campaign to create a "Coast to Coast Rock Highway"—the nation's first major highway devoted specifically to the needs of the automobile.

The two men put together an association to choose the route and raise money; the name Lincoln Highway was chosen (after some flirtation with naming the route after Thomas Jefferson), and by 1914 the group's brochures sported a new, soon to be familiar, phrase: "See America first."

The Lincoln Highway was to run first along the route of the old National Highway, from outside Washington to the Ohio, and then cross-country, along the course occupied today by Interstate 80, to San Francisco.

Fisher's notion was to build the road with contributions from the businesses that stood most to gain by the development of good roads. He and Joy obtained money from Goodyear. Local sponsors built individual sections of road: the association laid it out across the desert of Nevada and Utah. The organization aimed not just at creating this single road but at educating the public in the need for good roads, creating "seedling" miles to show the virtues of hard-surface roads to the locals.

The Lincoln Highway and its imitations marked the introduction of a fundamental element that would remain part of highway building from then on: boosterism.

No wonder the Lincoln Highway's boosters referred to it, as many another road would be referred to later, as "America's Main Street." It extended the values of the small-town Main Street and stretched them cross-country.

In 1920, Sinclair Lewis looked for the average American on Main Street. But by the end of the decade, he knew that he had moved to the highway. Lewis once advised a young writer to seek material for his novels by talking to the local gas station owner. In his 1928 novella, *The Man Who Knew Coolidge*, he has a character aptly named Lowell Schmaltz—a friend of George Babbitt—take off for Yellowstone Park. But before he can get more than a day down the road from Zenith he becomes so fascinated with the roadside stands— miniature Mount Vernons, wigwams, pagodas, and so forth—that he never makes it to the park.

The appeal of the Lincoln Highway lay in the fact that, even in the age of the railroad, the transcontinental trip still held an almost mythic attraction for Americans. Fisher's obvious model for the highway was the transcontinental railroad. Naming the highway for Lincoln, who as President approved the route for the railroad and pushed its construction as the ultimate symbol of the Union, was testimony to that fact. The building of the Lincoln Highway was a reenactment of the building of the Transcontinental Railroad, on new and more personal terms.

Every historian of the American automobile seems obligated to note that 1893, the year in which the Duryea brothers' first car appeared on the streets of Springfield, Massachusetts, was also the year in which Frederick Jackson Turner announced his famous theory of the influence of the frontier on American life—and proclaimed the closing of the frontier.

The automobile arrived in a world already settled by the railroad, which had opened and then closed the frontier. While it quickly took on the same air of historical inevitability enjoyed earlier by the railroad and later by the computer, the automobile from its beginnings was not associated so much with a vision of some modernistic future, but with the restoration of old values—the values that the railroad and big business had destroyed, the values of the frontier and the individual. It was seen through eyes of nostalgia, as a way to return to lost freedoms and a lost sense of the discovery of America. The car was from the beginning a tool of leisure and recreation, an invention for nostalgia. Sooner or later, it was assumed long before Henry Ford, everyone would have an automobile, probably an electric one.

This vision of the car offered a hopeful and reassuring response to Turner's notion that the frontier, the dominant force in American life, was now closed. The frontier's demands on the pioneer had tempered the American character, Turner had argued. Its open land and opportunity had served as "a safety valve" for social pressures. The very metaphor was one of the steam age: it required thinking of society as a great steam engine to imagine the need and function of a pressure valve. But the automobile promised to put the positive power of transportation in individual hands and offered an escape from the dominance of the railroad.

At first, however, the car figured largely as a toy, a means of recreation that could restore the freedom of the pioneering days.

Getting "off the beaten track" meant getting off the railroad track and behind the wheel.

With the arrival of the Model T, the automobile offered another hope: that of restoring the social mobility and economic opportunity of the frontier by tying country and town together. The automobile was quickly taken up as a means to keep open the frontier the railroad had closed. It held all the attractions of the railroad without the oppressive concentration of force. It figured as a sort of personal locomotive.

The first automobiles must have seemed almost as much of a luxury as private railroad cars. By the turn of the century, the automobile, usually in imported versions, was the star of garden parties among the Newport rich. Whitneys and Vanderbilts organized social events around rambles in their new cars. And the wealthy built the first roads specifically designed for the motorcar. In 1906 William Vanderbilt and some of his millionaire friends on Long Island's Gold Coast began construction of a concrete highway in which to exercise their vehicles. As car ownership spread beyond the very rich to the simply well-off, the car began its transformation into an object for mass leisure.

It took a long time for the auto to be associated with good roads. The early motor hobbyists seemed to glory in the difficulties of mudholes and other hazards. Motorists tried to outdo each other with tales of the hardships they had overcome on the roads. Putting up with "frontier conditions," mud, dust, insects, and breakdowns, was part of the game of the early auto rallies and caravans. For a long time there were real road hazards: sand and mud and ruts. Travelers carried planks and canvas to bridge the hazards. Auto touring held out the promise of reclaiming the sense of discovery that earlier generations—the pioneer generations—had enjoyed.

The first transcontinental auto crossing was made in 1903 by Dr. Horatio Nelson Jackson, a Vermont physician, who, with his mechanic Sewell K. Crocker, drove a two-cylinder Winton auto from San Francisco to New York. Despite nearly three weeks lost waiting for shipment of various spare parts, the trip took only 63 days.

Near Caldwell, Idaho, the party was joined by a stray bulldog named Bud, who spent the rest of the trip, like his human companions, wearing a pair of goggles.

The Lincoln Highway Association and American Automobile Association Guides were quick to admit—even to boast of—the difficulties of travel by car. A trip by car, the Lincoln Highway Association Guide noted in 1915, was like a "sporting trip." The mood of the road was that of Theodore Roosevelt's call for a return to "the strenuous life." Naturalist John Burroughs was one of the first to justify auto travel as a form of camping. Often, he said, we grow "wary of our luxuries and conveniences" and "long to get back for a time to first principles." The auto camper endured hardship "just to touch naked reality once more."

But almost immediately, efforts began to settle the auto frontier, to make it possible to rough it comfortably. Good roads became the cause, first of the hobbyist, the recreational motorist, and then of the farmer and would-be rural dweller. The farmer wanted to get to town; the urban middle class wanted to get out of town. The combination of the two forces produced a booster movement for improving highways.

The hardships motorists boasted of conquering offered a ready-made set of arguments for the Lincoln Highway. Fisher had a story that he repeated again and again to show the need for it. He and his companion had been driving one rainy, muddy night, he said, and came to a crossroads. They had no idea which fork to take, but there was a sign on a nearby telegraph pole, completely obscured with mud. Fisher got out of the car, scaled the pole, and rubbed the grit from the sign, only to find that it read: "Chew Battleaxe Plug."

Fisher's sales pitch even included a target date: the road was to be ready so that thousands of Association members could drive, in caravan, to the 1915 World's Fair in San Francisco.

Fisher raised some ten million dollars and obtained support from businesses, organizations, politicians, and hundreds of little towns eager to be included on the route. The Loyal Order of Moose of the World paved a mile of the highway near the group's head-quarters in Mooseheart, Illinois. California's progressive governor, Hiram Johnson, promised to pave his state's portion of the road without funds from the Association. Even President Woodrow Wilson, who once said that he feared the automobile would cause social unrest by creating jealousy among the working classes who could not afford cars, contributed $5. But Henry Ford, after initial enthusiasm, refused to join in. He remained more concerned, it seemed, with improving the back roads where owners and potential owners of his Model T lived.

Conflicts over the course of the road delayed its creation; every small town in America wanted to be on the highway, as once they would have wanted to be on the railroad. The breadth of national enthusiasm shrank decidedly after the final route was announced. For years, the "seedling miles" of the road remained connected only by dirt roads, and for much of its route "The Coast to Coast Rock Highway" remained little more than gravel.

Not until 1923, when the government stepped in, was the high-way actually finished. Almost immediately, the route was given the official designation of Route 30, the white federal shield and number replacing the local boosters' signs and the red, white, and blue colored blazes on telephone and telegraph poles. But the Lincoln Highway name persisted and persists in many places, along with those of the other roads local boosters created on its model: the Pike's Peak Ocean to Ocean Highway, the Yellowstone Trail, the Jefferson Highway (from Detroit to New Orleans). The National Parks Highway connected the parks of the West. (The Park Service since 1916 or so had been encouraging automobile use in the park in an effort to increase receipts from camping fees.)

Carl Fisher himself was one of the founders of the Dixie Highway Association, touting "your favorite route to and from Florida or the Great Smokies." Marked by red and white blazes on telephone poles, the Dixie Highway ran from Bay City, Michigan, through Chicago and Cincinnati, into Kentucky and Tennessee and around the peninsula of Florida in a great loop. It helped sustain the Florida land boom of the twenties—Fisher was also the prime mover behind the development of Miami Beach—and it boosted local sights along the way. Dixie Highway maps advertised Florence, Kentucky, as "the site of John Lloyd's famous novel *Stringtown on the Pike*," highlighted Jellico, Tennessee, as "the home of Grace Moore, star of movie, radio, and opera," and promised to show the traveler "the colorful, historic land of the old colonial homes and southern charm ... of genuine hospitality, of excellent food and modern, comfortable accommodations."

By the mid-twenties, the proliferation of Dixie Highways and Yellowstone Trails, many of them overlapping, each slashing its particular colored blazes across telephone poles, had become so confusing that many motorists had difficulty following the routes. Road indications—stop signs and danger signals—were still informal: a skull and crossbones for a dangerous intersection, a raised palm for a stop.

Rationalizing these signs was a task taken up not by the Federal Bureau of Roads but by AASHO—the American Association of State Highway Officials—the organization that was rapidly becoming the *de facto* ruling organization of road policy. Under the 1921 act that established the federal primary aid system—the U.S. routes—it was left to state officials to designate even the federal routes, since it was each state's prerogative to pick the roads—not more than seven percent of each state system—that were to receive federal funding and therefore the numerology of the federal shield signs. The shield sign literalized the "federal aegis." It was an emblem of the fact that the federal role was one not of construction or designation of highways but of protection and support.

AASHO was to grow in power with the growing importance of highways over the next half-century; it became one of the most important, least known political groups in the country. Founded in 1914, it was part lobby, part professional association, part quasi-political agency. No effective national highway policy could be enacted without its agreement. It was a grand version of all the local highway booster associations, a sort of Chamber of Commerce writ large.

In the late twenties AASHO produced its comprehensive plan

for road signs, creating or ratifying such familiar signs as the cross for railroad intersections and the octagonal stop sign. And later it was AASHO that formulated national standards for highway construction, embodied in its famous red and blue book for primary and secondary roads.

Ratified by the states and the Bureau of Public Roads, AASHO's standardized signs and numbering system marked an epoch in the rationalization of the highway, the shift from thinking of individual routes to thining of whole systems. As the auto trails, part of an exercise in frontier nostalgia, gave way to numbered routes, the frontier seemed to close once more. The neatly numbered map, whatever the varying condition of the physical roads, described a settled continent.

Roads became the dominant element of our maps, and road maps the most familiar kind of map. Physical features were downplayed in a road map; highways made them less significant. What was left was an amalgam of town names and road numbers. Never before had the average citizen needed any sort of map at all. He relied on informal directions or the services of professionals—coach drivers and others. But now, thanks to the automobile, he was venturing into unfamiliar terrain, where highways were constantly changing. The road map was the first democratic map.

In the days before the Interstates, the old *Saturday Evening Post* used to run a feature which showed a fragment of a road map, a circle like the view in a magnifying glass. The reader was supposed to guess the location of the circle. It was not easy: was "98" Texas 98 or Oregon 98? Was the town of Franklin the one in North Carolina or the one in New Jersey? The quiz, like the maps themselves, was testimony to the uniformity of the net the highways had cast over America, making near places strange and far places familiar.

Highways were necessarily depicted at far larger than accurate scale width on the maps that the petroleum companies gave away. (Often these maps would show each Esso or Philips station along the way.) And the standardized road signs provided a literal correspondence between map and physical reality.

•   •   •

The organizations to create the named highways eventually deteriorated into little more than extended chambers of commerce for
towns along the route. Their efforts were taken up by local booster
clubs like the Rotary and Kiwanis, for whom good roads became
a major subject of interest. By the twenties, the good roads movement had turned into a popular cause which the backwoods farmer
and the Main Street businessman could support with equal fervor.
It had become a major political force.

A rural Louisiana road (Monroe to Bastrop) before and after the Huey Long roadbuilding program, as illustrated in Long's campaign biography, Every Man a King. The politics of good roads dictated straight, simple highway slabs, built as quickly as possible. (LOUISIANA HIGHWAY COMMISSION)

3: *The Kingfish's Highways, or Everyman's Locomotive*

Along the course of Route 84, which runs across the Black River and Little Tensas Bayou and on past the cotton fields and scruffy pine forests of northern Louisiana, are posted thick metal signs with raised letters painted silver. They proclaim that this bridge or that stretch of road "was constructed during the administration of Govs. Huey Long and O.K. Allen."

More than half a century after those signs were erected, Long is remembered for having built one of the best state road systems in the country—and, better than any politician before him, having seized on the populist political reward of roadbuilding. (Allen was Long's trusted crony, head of his highway commission and successor as governor.)

Long's nickname was "the Kingfish," a deprecating term for a big shot he had happily adopted from the Amos 'n' Andy radio show. His political platform was "Share the Wealth." His fantastic slogan, "Every man a king," was also the title of his 1933 autobiography. The book is full of Long's country parables, his rantings against Standard Oil and the railroads, and his boasts about his administration's achievements. The frontispiece is a photograph of the famous Evangeline oak; beneath its limbs, Huey gave a celebrated speech in which he said that while Evangeline waited a long time under the tree for her lover, the citizens of Louisiana had waited longer for good schools, modern hospitals—and paved roads.

Long's boasts about the roads he built are accompanied by a whole series of before and after pictures: the "free" (no toll) bridge that carries Route 84 over the Black River near Jonesville, beside the ferry that was formerly the only way across; the Leesville to Alexandria road, just a washed-out swath of ruts surrounded by stark, straight pine trees in the before shot, a smooth asphalt ribbon post-Huey; Ponchartrain Road, sixteen feet wide, before, and built of asphalt at a cost to the state, the caption tells us, of $80,000 per mile, has become the Chief Menteur Highway, forty-one feet of "solid concrete," at a cost to the state of exactly half as much per mile.

Long was not a typical progressive or populist politician; he was not a typical politician of any type. But his career as a roadbuilder demonstrates, in paradigm and almost in caricature, the popular and political appeal that roads took on as the automobile became a staple of American life.

Part of Long's interest in roads was purely political and opportunistic, but part was rooted in a heartfelt if simplistic populist vision of the highway as a democratic common ground, where people of all classes and regions met on equal footing—like Steinbeck's characters in *The Grapes of Wrath* had.

Once, when he was governor, Long had his driver stop to pick up a woman hitchhiker and her two young children. He carried them to their destination and gave them money, all without revealing who he was. The incident summed up Long's populism—

the contradiction between the presence of the chauffeur and the gesture, between the showiness with which he later popularized the tale and the generosity.

Long's roads were built with an eye to their local political appeal. They were highly visible signs of progress, impossible to avoid, even for the man who did not own a car. Running into small towns like Winnfield, the roads demonstrated concern for "the little man," "the wool hat boys," "the fellows up the forks of the creeks."

His roads were every man's king's highways, but no one could ever forget that they were built by the Kingfish: the charismatic near-dictator, a parody of a monarch. Long was called a demagogue and a fascist. He was compared to Hitler and Mussolini, road-builders both, but he was actually more like one of the colorful American gangsters of his era: "the Caesar of the Bayous," in one newspaper phrase.

Long's roads were built as fast as possible. They were basic slabs of concrete or asphalt laid like carpet as straight as possible over the flat land, and over the low hills too, even when a slight curve would save a long grade. The right of way was simply a corridor through the cotton fields, or the unvigorous-looking pine woods that had grown up on the depleted soil of what once were cotton fields. There were few side ditches and virtually no shoulders.

Robert Penn Warren opens his novel inspired by Huey Long, *All the King's Men*, with a description of a road very similar to Route 84:

> You look up the highway and it is straight for miles, coming at you, with the black line down the center coming at you and at you, black and slick and tarry-shining against the white of the slab, and the heat dazzles up from the white slab so that only the black line is clear, coming at you with the whine of the tires, and if you don't quit staring at that line and don't take a few deep breaths and slap yourself hard on the back of the neck you'll hypnotize yourself and you'll come to just at the moment when the right front wheel hooks over into the black dirt shoulder off the slab, and you'll try to jerk

her back on but you can't because the slab is high like a curb and maybe you'll try to reach to turn off the ignition just as she starts the dive. But you won't make it, of course.

Route 84 runs west from the Mississippi through towns called Ferriday, Frogmore, Archie, Jena, and Tullos, until it reaches Winnfield, where Huey Long was born and where, hanging around the courthouse and the dry goods store, he first made his mark in the badinage and storytelling and intimidation that were to characterize his whole career.

Long was one of the first politicians to campaign via automobile: in 1918, running for the Louisiana Railroad commission on a platform of controlling the abuses of big business, archenemy of the "little man," he crisscrossed his district in a second-hand Overland 90, nailing up posters as he went. (Long's later campaigns were virtual automotive blitzkriegs, featuring cars equipped with mobile speakers that his chauffeur regularly drove from town to town at 70 or 80 miles an hour, on one occasion wearing out six sets of tires in a single week.)

In this Long violated a political maxim that held that country people would resent as pretentious someone who campaigned in a car. But Long, a former traveling salesman, knew the esteem in which the automobile was now held in the country. When he ran for governor in 1928 he carried the same realization further and made roadbuilding a key part of his platform.

When Long took office, there were fewer than fifty miles of concrete road in the state and about seventy-five miles of asphalt road. The state highway program was $5 million in debt. To make his road program possible, Long pushed through constitutional amendments and a huge road bond issue that passed the required referendum by margins of twenty to one. On a map of the state he drew the places the roads should go. His map was a series of short sections of road centered on parish seats. As a system, Long's map made no sense. Its purpose was to deliver some roads as soon as possible to as much of the state as possible—and to whet the local appetite for more roads. In private, he projected an extensive and complete network of modern highways for the state that would join all those short segments together.

North Carolina had model roads. In the first half of the twenties, under Governor Cameron Morrison, the state had built some 6,000 miles of improved roads, using, as Long would, bonds and gasoline taxes to finance the construction. Long hired the chief engineer of North Carolina's road system, Leslie R. Ames, and twenty-one other road administrators and engineers—virtually the entire North Carolina highway department. Ames's North Carolina salary of $6,000 was raised to $13,000, the highest of any state official.

By 1931 Long had built more than 2,000 miles of surfaced roads. The road program had also turned into a relief program to ease the effects of the Depression. Louisiana employed 22,000 men on road-building, more than any other state—Franklin Roosevelt's New York was second—and two-thirds of the state budget was going to the program. Long's program anticipated the roadbuilding projects of the Works Progress Administration. His rise as a potential threat to FDR in 1936, along with the "Roosevelt recession" of 1937, were responsible for pushing the President toward elaborate plans for new roads.

Long pioneered some of the modern methods of highway financing, although their principles had been established at least as early as the Gallatin report. He created a whole financial machine for roadbuilding, issuing bonds secured by gasoline and other taxes and diverting the revenues from the stiff taxes he had imposed on his archenemy, Standard Oil, to roads.

The rest of this roadbuilding machine was the political product: patronage and contracts, which could in turn be used to twist legislative arms for further road authorizations.

Roads, of course, had always been paved by the pork barrel. "The highways of America," Carl Fisher had lamented while boosting the Lincoln Highway, "are built chiefly of politics whereas the proper material is crushed rock or concrete." Gallatin himself had written Jefferson, during the planning of the route for the National Road, that strategic Washington County, Pennsylvania, "gives a uniform majority of about 2,000 votes in our favor, and if this be thrown by reason of this road in a wrong scale, we will infallibly lose the State of Pennsylvania at the next election." The National Road went through Washington County.

Long rewarded supporters with lucrative concrete and shell contracts. (Shell was a cheaper aggregate for concrete than gravel in many parts of the state.) His political machine received large contributions from road contractors, and he used roads as a source of patronage and contracts for his supporters. (One contractor, who was also a state representative, was convicted of income tax evasion in connection with a sweetheart deal to provide shell for state highways.) Despite creation of a supposedly impartial advisory board, Long retained virtually total control over where the highways would go. People in Louisiana, however, were used to corruption. In Winn County it was acknowledged that while Huey probably stole a dollar for every dollar he spent on the roads, his predecessors would have kept both dollars.

The popular pressure for better roads that Huey Long exploited had been growing steadily for a generation. The Good Roads movement predated the Lincoln and Dixie highways, even predated the automobile. Its first champions were bicyclers, whose organization, the League of American Wheelmen, held in the 1880s and 1890s something of the status of the Sierra Club today. By the 1890s there were some ten million bicycles in the United States.

It took more than the arrival of the popular automobile, marked by the introduction of the Model T in 1908, to establish the fact that the automobile would require new types of roads: it took the establishment of travel by car as a commonplace part of daily life.

The Model T was a car whose virtues lay not only in low price, durability, and ease of maintenance and improvement, but in the fact that it could deal with the existing roads of rural America, wagon roads, carriage roads, unimproved roads. Between its introduction in 1908 and the end of production twenty years later, some 15 million were sold. Auto sales and the pressure to improve roads fed off each other.

Long before Henry Ford, it had been widely assumed that the

automobile would become a universal item. The questions were when, and who would make it happen? Sears even sold its own automobile, by catalog, between 1903 and 1910. By 1920, an automobile could be bought for as little as $300. And the coming of installment buying and a market for trade in used cars made owning a car even more accessible. Soon, more than half the new autos in the country were bought on time. The development of installment buying helped triple car ownership between 1920 and 1930. By 1925, half the families in America owned cars.

More and more cars were of the closed variety—some eighty percent of new cars by the mid-twenties—cars that could keep their passengers dry in any weather. Demand grew for roads that also could function in any weather. Good roads finally killed the Model T: more modern vehicles offered greater comfort on the better roads.

Nothing better showed how the car demanded new surfacing than the opening of the famed Indianapolis Speedway, on whose original mud surface, not much different from a horse track, just one race was run before the promoters saw the necessity for hard pavement. The famous "brickyard" was the result.

In a few places, like forward-looking Bellefontaine, Ohio, streets were being paved with concrete of cement or asphalt base. And in 1908 Wayne County, Michigan, began planning the nation's first system of concrete highways. For most of the country, however, the goal of good roads simply meant improving rural roads to minimal standards. At the most, it meant gravelling; at the least, grading dirt roads to remove the mudholes in which cars became mired, and reducing the steepest grades. The age of the railroads had seen American roads decline to their worst state ever. It was estimated that in 1900 the state of Iowa had more roads with excessive slopes than all of Switzerland.

The push for better roads took on force with its links to the various rural progressive movements—from the Grange to the Progressive Party—and the arrival of the automobile. Better roads were a weapon against the railroad, which was widely perceived as the root of the farmer's problem: the railroads and their continued high freight prices in the face of falling crop prices. Good roads meant easier access to competitive railheads and local farm markets.

It was the railroads that inflated prices for carrying farmers' crops to market while granting rebates to large corporations like Standard Oil and U.S. Steel. It was railroads that had consumed huge amounts of public land and treasure and diverted them to corruption. It was the railroads on whose timetables a man had to wait. It was the railroads that formed the "Octopus" of Frank Norris's novel, which seemed to have come about by themselves and which no one could control.

Sensitivity to the inherently undemocratic nature of the railroads began early, even as the country was in the first blush of its love affair with the locomotive. In an 1847 speech at the dedication of a new railroad, Daniel Webster attempted to refute "idle prejudices" against the closed nature of the railroad corporation. "The track of a railroad cannot be a road upon which every man may drive his own carriage," he argued. But within fifty years, the automobile had made just such a road possible, and assumed its place as rail's natural, democratic rival.

The railroads represented the abuses of big business and complaisant government at their worst. They were also the most convenient villain for rural Americans since Andrew Jackson fought the Bank.

The great era of railroad building, epitomized by the transcontinental railway, the space program of its time, valued rail as a means to "bind together the Union." In a sense, it represented a continuation of the Civil War by other means.

There was a neat irony to the fact that General Sherman, whose famous March had invented the logistics of modern war and who had systematically torn up Southern railroads en route (rails, heated white hot and pretzeled around trees, were known as "Sherman's hairpins") later directed the construction of the western railroad. The mad dash to complete the link led to the employment of military means of organizing labor and materials.

Many Union generals were hired as front men, capital raisers, or, like Sherman, executives of railroads. U.S. Grant was called in shortly before his election as President to make peace between rivals for control of the Union Pacific railroad.

The transcontinental railroad was linked to another achievement of the Lincoln Administration: the Homestead Act. Free land for the railroad companies was to be balanced by free land for settlers in the West, an attempt to realize the Jeffersonian—and anti-urban—virtues of land ownership.

The place of both measures in sketching out an expansionist future—better transportation to open up agricultural lands and industrial markets, free labor as opposed to slave—in the debates of the Civil War issues was important. Not until the slavery question was resolved could the nation agree on the course for the railroad or the status of the new territories.

The Homestead Act, like the land laws of 1785 and 1796, failed to put land in the hands of small farmers. The best land went to speculators or to the railroads themselves, who might resell it for huge profits. Dry and unproductive land was often foisted off on unsuspecting immigrants, lured by flyers distributed throughout the East and even in Europe and transported west at bargain rates.

And the railroads, rather than Jefferson's ideal grid, set the pattern for town development that the automobile would follow. Towns sprang up along the construction route of the transcontinental railroad; some took root, some vanished when the hostlers, whores, and the camp followers—in many cases the same entourage that followed the Union armies from camp to camp during the war—moved on.

There were hundreds of such towns, built along the route of the railroad, called Hells on Wheels for their morality and means of arrival. "Here's Julesburg!" cried a Union Pacific freightman, as his train pulled up to a barren site in Wyoming. The entire town was a set of flats, a virtual stage set, that came in on rail cars. The speed and style of railside construction was to influence the creation of the roadside as much as the intensely organized construction methods of rail were to inspire the "paving trains" of modern highway building.

• • •

Good roads were thought to be an ally of the small farmer. Roads and the individual mobility they provided would restore the freedom of the frontier, the dignity of the small farm and the husbandman. To be a good roads man was to be considered a solid progressive, possessing a vision lacked by the "Get a horse!" sceptics.

"Get the farmer out of the mud" was the catchphrase, as if the automobile would somehow make farming as clean as accounting. But mud was also a physical problem; in his traveling salesman days Huey Long learned the common trick of carrying around a bottle of moonshine to buy the help of a local farmer and his team in case his car got stuck on some muddy back road.

In fact, the thing the farmer liked best about the car was that it would take him to town. No longer was it a choice between plowing the south forty or saving the mule for the trip to the general store. The T, or a Fordson tractor, could plow, and the T could carry the family to the moving picture show the same night. Thus the importance of the rural road. Justify it as they might as "farm to market," the improved rural road was also a route to town culture for people who, if asked, would nonetheless obstinately insist on the primacy of rural values over urban.

When William Jennings Bryan, the champion of the Populists and the first man to campaign for President from an automobile, was to speak in some rural location, often a fairgrounds or revival site, it was said that he was "always good for forty acres of Model T's." (Long claimed to have borrowed his "every man a king" phrase from Bryan, who spoke of "a Republic ... whose every man is a King, but no one wears a crown.")

Bryan might divert and entertain the farmer on a Saturday afternoon, haranguing them, as he did in the famous Cross of Gold speech, about how quickly grass would grow in the streets of cities without the farmers. But the next Saturday it was just as likely as not that half those T's would be parked downtown in the vicinity of Main Street. And those cars, with better roads, would take them even further, to the city, and perhaps eventually even to California and Florida. By the end of his life, Bryan would be using his silver tongue to tout real estate in Coral Gables. The promoters knew their audience.

Already the automobile was helping to bring the benefits of the big city to the country. Rural Free Delivery, which demanded the improvement by local authorities of thousands of miles of rural roads, became a key plank of the Progressive movement, and its approval in 1896 immediately accelerated the rise of firms like Sears Roebuck and Montgomery Ward, whose goods began to filter into out of the way farms in Louisiana and Iowa.

While increasing the pressure for additional state highway building and for the federal government to begin highway aid, RFD also inspired farmers to improve their roads themselves. Thus was born one of the unsung inventions of the twentieth century: the King drag.

Invented in 1905 by an enterprising Maitland, Missouri, farmer named D. Ward King, the King drag was nothing more than a log split in half. The halves were joined by scrap boards a couple of feet apart and the front half was edged with a metal scraper. Attached to a team of horses or mules, the drag smoothed away ruts on thousands of miles of rural dirt roads, and was adopted by the Federal Bureau of Public Roads as a device to bring back roads up to RFD standards. By 1916, RFD and improved general postal service had given Woodrow Wilson's administration the constitutional justification it sought to begin federal aid to state highway building programs.

Good roads had become a symbol not only of the possibility of greater physical movement, but of upward class movement. The popular automobile became a sign of the breaking down of old class barriers. In particular, it became an agent for the liberation of the "little man," as he might be called, and, in the South especially, the redneck, white trash who were shiftless, moved about aimlessly to squat on one piece of land until they were driven off to find another—Faulkner's Snopes.

This process is the backdrop for the changing world of Faulkner and other southern and midwestern writers. The car literally brings about the death of one of Faulkner's last aristocrats, old Colonel Bayard Sartoris, who suffers a heart attack when his grandson drives it off the road. And the Snopes, the white trash clan who infiltrate Yoknapatawpha county, move from mules to used Model T's after World War I. The Snopes are particular patrons of RFD: Flem Snopes names one of his sons "Montgomery Ward."

Hawthorne, in "The Celestial Railroad," a sort of parody of *Pilgrim's Progress*, had depicted railroad culture as a false god, an illusion of salvation. The railroad presents Everyman with the temptation of a free and easy ride to heaven; in fact, of course, it turns out to be an agent of the devil and its destination to be damnation.

The automobile, however, put Everyman in the driver's seat and let him choose his own destination. With good roads, the car would become Webster's imaginary railroad where every man can drive his own carriage. And the social and moral implications of that system were revolutionary.

"No man with a good car needs to be justified," says Hazel Motes, the deranged evangelist of "The Church Without Christ" in Flannery O'Connor's *Wise Blood*. Here in brief was the impact of the popular car on rural America and the Bible Belt, touching with that word, "justified," the whole Protestant system. Driving yourself was the very essence of the Protestant ethic, as the fundamentalists knew well who nailed their "Get Right With God" signs on the pine trees beside the road. The car presented a danger, a temptation of its own, and it was a risk that each driver ran alone.

The political side of this dangerous innovation was clear to some— Woodrow Wilson had feared that the car would make the poor jealous of the rich, but that was before Henry Ford. To many more, the automobile offered a healthy balance to the challenges of godless ideology. Socialism and communism, by the end of World War I and the Red Scare of the early twenties, were already perceived as threats.

The political consequences of the replacement of the railroad by the automobile were clear to the Chevrolet copywriters who created a 1924 magazine ad including these words:

> The automobile, 14,000,000 strong, has in truth become our most numerous "common carrier."
>
> Every owner [is] in effect a railroad president, operating individually on an elective schedule, over highways built and maintained chiefly at the expense of himself and his fellow motorists.
>
> What has been the effect of the automobile on our composite national mind?—on our social, political, and economic outlook?
>
> The once poor laborer and mechanic now drives to the building operation or construction job in his own car. He is now a capitalist—the owner of a taxable asset. . . .
>
> Evenings and Sundays he takes his family into the country or to the now near town fifty to one hundred miles away. He has become somebody, has a broader and more tolerant view of the one-time cartoon hayseed and the fat-cigared plutocrat.
>
> How can Bolshevism flourish in a motorized country having a standard of living and thinking too high to permit the existence of an ignorant, narrow, peasant majority?

With the automobile each could take his own road, if not necessarily to Heaven or Hell, at least to town. But just how far he could go, of course, depended on the roads.

Good roads became mixed with Babbittry. Good roads would build up Main Street, for wasn't that where everyone was trying to get? Eventually, it was argued, good roads would turn small towns into new Chicagos. It was the same false promise that had been held out at Julesburg.

As the automobile came to be seen as a piece of individual capital, it became necessary to individualize automobiles. The competitive answer to Henry Ford's success in making all his cars the same was Alfred Sloan's decision to make cars as different as possible: to produce a full and varied "line" of Chevrolets, Buicks, Oldsmobiles, and so forth under the General Motors aegis and to change their look each year. In 1927 Sloan took a fateful step and hired the first automobile stylist, Harley Earl. Sloan had recognized that the automobile had now become akin to a suit of clothing, indicative of how wealthy and how "up-to-date" the driver was. The impression you made on the road was as important as the one you made on the street. The comparison to clothing was not an idle one. With the development in 1924 of Duco lacquer, Model-T black gave way to a rainbow of colors. Mixing and matching different brands, models, bodies, power plants, colors, and optional equipment, the automobile consumer would eventually be able to choose among potential variants numbering in the hundreds of thousands each year.

Sloan tied the desire for upward mobility to installment buying to produce a system where the ideal buyer worked his way up the GM ladder, trading his first Chevy for a Buick or Oldsmobile and ultimately aspiring to the Cadillac.

This ladder of automotive status created the supply of used cars, sold so cheaply almost anyone could buy one. Now Everyman had a car, however ancient or beat-up, and to the rest of the world that made him seem like a king. When the film *The Grapes of Wrath* was shown outside the United States, foreign viewers were confused. To them, the idea that the Okies were poor was laughable: didn't they drive their own cars?

The soil was never very good in Winn County, and cotton farmers existed at a minimal level until the fields finally gave out altogether and were abandoned. The railroad came in in 1901, when Huey was growing up, bringing the lumber companies with it, and the area enjoyed a burst of prosperity that swelled its population to nearly three thousand before it was "logged out." The straight pine forests along Route 84 have grown up since.

Today beside 84 there are still cotton fields, long and wide, bright green in spring and summer when hot mirages make the asphalt seem to melt into liquid tar, miraculously rich brown in the fall when the bolls break open and the white of the fibers sings out against the dead plants. But between the planted fields are more and more abandoned ones, overgrown with blackberry, sumac, and scrub pine.

At regular intervals you encounter convenience stores, most of them brand new, replacing old country stores with rusting roofs of corrugated steel that Walker Evans might have photographed. "Beer Milk Bread," they advertise, plus gasoline and Doritos and Moon Pies and Sprite.

Follow the road into Winnfield today, past the Huey Long Motel, on the edge of town, where through a window in what is apparently the only occupied room you see a man reach up to adjust his television set—you will see how Huey's roads have changed it.

Downtown there is an ugly courthouse—singularly ugly, considering the classical virtues of many seats in surrounding countries—a Western Auto doing poorly, the old dry goods and clothing stores just getting by and two film theaters—the Venus and the Palace—closed up for good. Downtown Winnfield is like one of the depleted cotton fields out in the county, abandoned by farmers and plantation owners who have moved west to fresher fields, the stone buildings cold and barren as weak soil.

But long after what little activity there still is downtown has ended, the lights glow warm and open on the strip of 84, northwest of town, with its shoping centers and fast food and more convenience stores. This is a phenomenon of twentieth century road culture as neatly traceable as the natural succession of forests: like a hundred thousand other small American towns, Winnfield long ago moved out to the highway.

The Taconic State Parkway. The height of the parkway art, recapitulating the ideal of the English country landscape.

4: *Parkway to Freeway*

In *The Great Gatsby*, F. Scott Fitzgerald described the huge painted eyes of an oculist's advertisement—for one Dr. T.J. Eckleburg— and the "valley of ashes" behind it. Whatever larger significances they took on in the novel, these roadside sights were based on real ones of the twenties. There was such a sign, and the ashheap was the Corona dump, beside the road to New York from Great Neck, the suburb where Fitzgerald himself had lived and which in his novel he transformed into East Egg.

In the novel, congestion and roadside distractions make the road a dangerous place: an automobile accident brings about Gatsby's undoing. It is while waiting at that special villain of twenties and thirties traffic planners, the at-grade railroad crossing, that Nick Carraway studies Dr. Eckleburg. The crossing was a symbol of the passage of transportation dominance from the railroad to the automobile highway. Planners now sought to remove these crossings. Cars could no longer be asked to wait for the train, and drivers could not be asked to risk the dangers of stalling on such crossings.

Had Fitzgerald or his characters made the same trip ten or fifteen years later, they would have seen the rail crossing spanned by a bridge, and the old narrow road replaced by a divided highway, Robert Moses's Grand Central Parkway. And on the old site of the valley of ashes, they would have seen the buildings of the 1939 World's Fair, housing visions of even grander highways.

The stretch of road Gatsby drove showed the problems of the auto and the move out of town it had brought. In the first decades of the automobile, trouble started when you got out of town, left the pavement and hit dirt country roads. Now, the slowdown started when you reached the edge of town, with its roadside congestion.

Roadside ugliness and danger went hand in hand. The roadside eatery or billboard was a distraction to the driver, and the cars entering the road from parking lots were a traffic hazard. Fatalities and injuries from auto accidents rose more rapidly even than the number of automobiles. In 1928 some 28,000 Americans died on the road. As measured by fatalities per miles driven, the twenties were the deadliest highway decade ever: eighteen people died per million miles driven in the late twenties; in 1960 it was only five, and, except for a jump in the late sixties, the figure has continued to decline.

The spread of billboards had been decried since the turn of the century. The sprawl of the roadside, planless, crude, vivid, and exciting, distracted the driver and wore on his senses. It was shrill and static-ridden, creating in the driver what would later be known as sensory overload.

In the early thirties, a forty-eight-mile strip of U.S. 1 was found to have nearly 3,000 buildings with direct access to the road. There was a gas station every 895 feet.

In New Jersey, U.S. 1 was even worse. It had 700 businesses and 472 billboards along the forty-seven miles between Newark and Trenton. Margaret Bourke-White, dispatched by *Life* magazine to photograph the side of U.S. 1 between New York and Washington, found it "one long clutter of ugly signs."

"'The nation that lives on wheels'" lamented *Life*, "still has the dubious honor of having created, along 3,000,000 miles of highway, the supreme Honky-Tonk of all time." Roadside clutter was particularly offensive to those who, increasingly, were taking advantage of the automobile to move to the suburbs—to all those who aspired to the lifestyle of a Gatsby. And it was clear to all that the suburban revolution begun by the commuter trolley and railroad had entered a new, accelerated stage thanks to the automobile.

This trend was clear. The President's Committee on Recent Social Trends, appointed by Herbert Hoover, reported in 1932 that the automobile "had erased the boundaries which formerly separated urban from rural territory and has introduced a type of local community without precedent in history."

To smooth the way to the new suburbs, to ease congestion, make driving safer, and remove unsightly roadsides, planners began to dream of new sorts of auto highways, differing more from the two-lane blacktop than that road differed from the muddy carriage track.

The Grand Central and the other parkways of Long Island and Westchester County were created in response to the ugliness and congestion of strip development.

The term *parkway* was coined by Calvert Vaux, collaborator of Frederick Law Olmsted in the creation of Central, Prospect, and other parks, in an 1868 proposal that led to the creation of Ocean and Eastern parkways in Brooklyn, the first parkways in the world. The models for these parkways were Haussmann's Parisian boulevards, for example, the Avenue Foch, Unter den Linden in Berlin, and the drives in the Bois de Boulogne in Paris, which Olmsted admired greatly.

Olmsted and Vaux found other precedents in the divided streets of colonial towns and their extension into such streets as Commonwealth Avenue in Boston, laid out in 1858. In these models, the principles of design with an eye to the pleasant carriage ride, the separation of pedestrian and vehicular traffic, and the use of a central median to divide traffic in each direction were well established by the first half of the nineteenth century. Olmsted and Vaux added strips of flanking park, with benches and patches of "greensward," and placed regulations on development along their margins.

The parkway was a natural successor to Olmsted's romantic view of the park as a series of sequential vistas, with its basis in the parks of English country estates—and in the groomed landscape of New England where he grew up. Olmsted valued a succession of views to the pedestrian or carriage, with vistas opening up suddenly, alternating with cozy, close forests; there was to be great variation in landscape, not natural but manipulated by the hand of man and, ideally, representing a variety of different landscapes, a miniature grand tour, a local anthology of picturesque landscapes sought out by the traveler.

As Haussmann's designs were conditioned by the Revolution of 1848—his boulevards allowed for the speedy arrival of police and firefighting forces at any imminent riot, and the paving stones that were the ammunition of the republican uprisings were replaced by solid pavement—Olmsted's parks and parkways were conditioned, very differently, by the Draft Riots of 1863, in which class and ethnic discontents found their focus in the conscription laws, with a loss of life comparable to the worst Civil War battles.

Olmsted's answer to such conflicts was to provide a recreational safety valve, and to place classes and ethnic groups in proximity in his parks. Proximity, but not necessarily contact: the separation of carriage traffic and pedestrian traffic could also be viewed as a separation of the upper and lower classes, although the aristocratic promenader was common as well. Olmsted saw his parks, with perhaps as much romanticism as he injected into their bow bridges and rambles, as institutions for democratization. His parkways had the same function, linking parks, making drives through whole cities possible, but above all tying the city to the suburb, where, ideally, the urban and the pastoral would find their happy equilibrium in a middle class. The parkways were to serve to foster and direct this sort of development in Brooklyn, and later in other cities—Boston, Buffalo, and Chicago.

For Olmsted the parkway was a way to gain more land for parks while providing transportation. But the parkway evolved into the idea of the ring park, or greenbelt, and thence to the visions of Benton MacKaye, whose 1920s notion of the roadless city, cityless road turned the parkway into a suburb. "Like the fly, the motorist buzzes his wings vigorously, but his feet are stuck to the flypaper of the old-fashioned highway," MacKaye wrote, where "a spavined horse could travel as fast as a 120 hp car."

As the axiom from which all highway design should derive, MacKaye picked up on the notion of the car as a personalized form of railroad; as he put it, the automobile should be thought of not as a horseless carriage but "as the family locomotive," for which new types of highway, as distinctive as tracks, should be built. He argued for wide highways in parkland that bypassed towns, and banned "roadtown," as he called strip development.

No one ever talked about beautifying the side of the railroad; the railside was private property. But the roadside, at least as far as the right of way, was public; and with the automobile, looking out the window was more important: it was unavoidable.

It is true that the first fancy motor roads began as private institutions.

The nation's first public motor parkway, the Bronx River Parkway, begun in 1907, was created in response to the deterioration of the Bronx River landscape: ducks in the Bronx Zoo were dying in the ponds fed by its waters, and shacks and advertisements lined the banks of the polluted river.

The Bronx River Parkway, which ran from the Zoo to the Kensico Reservoir, was a divided highway with a narrow median that passed under crossroads via a set of overpasses. Some 30,000 trees were planted along its margins. It was a recreational highway designed for speeds no higher than 35 miles per hour. Commercial vehicles were banned. It inspired a whole series of other parkways in the area: the Saw Mill River, the Hutchinson, the Cross County. Most followed the courses of small rivers. The system shaped the suburban patterns of Westchester County. Among the backers of the Bronx River Parkway were a number of Westchester County real estate men: the Parkway helped create Scarsdale. A broad, screened right of way and the limitation of access enhanced the value of abutting property, the planners pointed out. By 1932 Westchester County had completed 160 miles of parkways.

The parkways of Robert Moses, mostly designed by Clarke and Rapuano and Clarence Combs, extended these design principles to Long Island. Now it was the middle class—which could be neatly defined as the automobile-owning class—for whom the parkways were intended and designed. (Robert Caro recounts how Moses intentionally had the parkway overpasses designed too low for buses, in order to keep out the unautoed and presumably unwashed—the new immigrants, in particular—to say nothing of his refusal to consider rail links to the parks at the ends of his parkways.)

Moses's Meadowbrook Parkway appears so fresh and modern, even in the 1980s, that the odologist finds it hard to believe that it was opened in 1934, a year before the first *Autobahn*.

With its weathered-wood lamp standards and guardrails, its wide, gently sloped shoulders, the half-artful, half-natural grouping of the plants, and the stone facing of the arched overpasses, the Meadowbrook blends brilliantly with the low forest around it. Entering the wetlands, it crosses bridges with wonderful little Deco guard stations and finally sweeps in great loops, among reeds swept by seabirds, onto the Long Island beaches.

The culmination of parkway development was the Taconic Parkway, running north from Westchester County into the countryside of the Taconic River Valley. The Taconic was every driver's vision of the ideal track for a Sunday drive. Its roadways are widely divided by a meadowed median, spotted with wildflowers and often punctuated with copses of trees that could have been taken from a Constable painting, sometimes thickening to a full hardwood forest that reaches practically to the edge of the rustic guardrails.

Begun in 1940 and completed only in 1950, the Taconic is one of the most praised highways in America. It is a parkway with the best of that form's landscaping traditions, but without the tight and sometimes dangerous curves of parkways built for slower traffic.

Lewis Mumford, a severe critic of most highways, liked the Taconic. He called it a "consummate work of art" and praised its "opening up great views across country, enhanced by a lavish planting of flowering bushes along the borders."

The Meadowbrook and Taconic parkways marked the apogee of

the parkway tradition: expensive, elegant highways for affluent suburbs, gardens for machines. It was no wonder that, with their wide, grassy medians, curving vistas, and neat clusters of plantings grading into rough grass and brush at the edges, the parkways suggested golf courses, with fairways and roughs.

The federal government seized on the parkway idea as a way to create jobs during the Depression and to bring the automobile into the national parks and forests. The pioneer federal project was the Mount Vernon Memorial Parkway, begun in the late twenties. In the Blue Ridge Parkway and Skyline Drive, in the Norris Freeway built by the TVA, the parkway functioned as a way to show off great landscape. Along the Natchez Trace Parkway, the landscape was tied to the history of the area: old portions of the original Trace, along with cabins and an inn, were preserved within its right of way.

The parkway would eventually seem most important as a forerunner, pioneering the legality of limited access and the divided roadway. The postwar year saw the construction of more parkways—the Palisades Parkway, the Garden State Parkway, and others—but they were distinguished from other highways mostly by their earthtone "decor," by the wooden browns and stone grays the parkway builders loved: the wooden lamp standards, the guardrails of creosoted beams (or, later, of naturally weathering steel alloys), the grouped plantings in the median, the arched bridges faced in local stone, even the directional signs with their brown backgrounds.

But the highway designed for beauty and recreation was giving way to the overridding necessity of the flow of the road. E.M. Bassett, the lawyer who is credited with coining the term *freeway*, contrasted it with parkway. The parkway, he said, was "a strip of land dedicated to *recreation*, over which the abutting owner has no right of light, air, or access." The freeway, by contrast, was the same sort of strip "dedicated to *movement*."

A superhighway could also be a superparkway. In 1943, the

Highway Research Board began touting its idea of "the complete highway." "The same basic qualities," its report argued—"utility, safety, beauty, and economy"—"should be integrated into the design and construction of every road, whether it be local, county, state, interregional, or international."

California's future role as the state most associated with the freeway was founded on its early passage of a state limited access law in 1939. A year later, in December of 1940, with the Rose Parade Queen and her court in attendance, and balloons in abundance, California Governor Culbert Olsen snipped the ribbon and the Arroyo Seco Parkway in Los Angeles was opened. By the end of the war, the Arroyo Seco had become the Pasadena Freeway and the parkway concept had been subsumed in that of the freeway.

Beside the Grand Central Parkway, on the site of the old garbage dump, the 1939 World's Fair offered grander visions still of new highways, superhighways designed specifically for the modern motorcar. Hitler's Autobahns were widely reported in the press, and at General Motors's Futurama exhibit, designed by streamline designer Norman Bel Geddes, visitors could see a vision of the country crisscrossed by such superhighways, wide, sweeping, twelve-lane highways that leaped across canyons on slim bridges and met in huge, ramped intersections.

Modern cars had become faster and faster. While the Model T strained to hit fifty, a 1935 Ford could cruise at seventy-five. The limits, now, were in the highways. Futurama showed cars humming along at 120 miles an hour in the fastest lane, on the left. Laggards piddled along at 75 in the right lane. Crossing the country on such highways could be done in a day's driving. Even more significantly, Bel Geddes claimed that with his new highways commuting distances could be increased sixfold.

There was more to all this than futuristic speculation. The Futurama exhibit was designed to fire the public imagination with the possibilities of superroads. It was a device for education and propaganda. The auto, oil, construction, and other road-related industries were moving to support a major national roadbuilding program that would foster sales of their products.

After stepping down as General Motors' chief operating officer in 1937 to become chairman of the board, Alfred P. Sloan, the dapper administrative genius who had built GM into the world's largest manufacturer, organized a lobby called the American Highway Users Association.

In October of 1939, he held a dinner at the Futurama to court the state highway officials of AASHO and help sell them on the vision of magic motorways. The officials were swept through Bel Geddes's dream highways and then gathered together to contemplate the possibilities of realizing some of them. Sloan's company, naturally, could benefit from such a system. It represented an aggressive response to contentions that the auto market was saturated.

Other industries were already involved in the coalition of interests that would dominate postwar highway construction. The American Institute of Steel Construction touted elevated highways. The Portland Cement Association argued that a $6 billion roadbuilding system would more than pay for itself in economic growth. Such groups would invite Bel Geddes in to show his models and give his speech on magic motorways.

Franklin Roosevelt, too, was intrigued with the idea of the superroads. He invited Bel Geddes to the White House for a private "stag" dinner on March 22, 1939. With his promotional eye, Bel Geddes brought along a model of a streamlined yacht he had designed, a toy sure to catch the fancy of the nautically minded former Assistant Secretary of the Navy, and a devoted sailor. Bel Geddes's highway ideas also made an impression. Bel Geddes correctly diagnosed the conflicts, among federal, state, and local authorities as a major part of the highway problem. He argued for a national highway planning authority and a fifty-year construction plan. FDR, however, never allowed himself to be publicly associated with such a dreamer as Bel Geddes. His magic motorways were good publicity, but still a bit farfetched.

Roosevelt had his own plan. He had a longstanding interest in highways. Before becoming governor of New York, he was chairman of the Taconic River Commission, which was planning the Taconic Parkway. It was a pet project of his; he even worked on the design for picnic tables along the route and considered the proper type of stone to face the crossing bridges.

Roosevelt had watched Huey Long's progress in Louisiana. In the months just before his death, Long seemed Roosevelt's strongest potential challenger for the 1936 election. The appeal of Long's highway-building program, tied to his notions of income redistribution, were clear.

By 1942 the Roosevelt administration had spent some $4 billion on highways, and since 1937 he had been exploring the idea of a major national superhighway program. Early that year Roosevelt called in Bureau of Public Roads boss Thomas McDonald, the legendary "Chief," and handed him a map on which were drawn six lines, three running north to south, three east to west.

To build these superroads, Roosevelt envisioned a national land authority, with right of eminent domain, which would take over land along the route of the highways and pay off its capital and interest by gradually reselling the land to local government or private business.

The railroad experience of vast and abusive private profits enjoyed from government land grants was clearly in the background of Roosevelt's thinking. "Why," he argued, "should the hazard of engineering give one private citizen an enormous profit? If there is to be an unearned profit, why should it not accrue to the government, state or federal, or both?" To most of the country, this smacked of socialism. In 1938, Congress had already rejected an ambitious $8 million road plan.

McDonald and his staff came back with an answer to Roosevelt. A Bureau of Public Roads report released in January of 1940 estimated that a realistic 29,300-mile superhighway system would cost $6 billion. A minimal 14,336-mile system—Roosevelt's six roads—would cost $2,899,000,000.

There was not enough traffic to justify such a plan, the report argued. The BPR was probably also sceptical of methods FDR suggested. Toll roads represented a threat to its power, and, since most of them would be administered by completely new, independent agencies, to that of the AASHO engineers from whom the BPR really derived its power.

In its report *Toll Roads and Free Roads*, the Bureau of Public Roads argued instead for an "interregional system" of less ambitious, toll-free superhighways. The BPR's scheme was built up from existing and projected traffic patterns. The bureau created a map showing traffic flow in a pattern of peaks and plains around the country, with a huge range of traffic along the east coast, topping out in the Northeast, and great flat stretches where Roosevelt had wanted the transcontinental routes to run.

The BPR had no notion that the construction of new superhighways, like the introduction of such inventions as the telephone and the auto itself, might create its own demand. Within a year after *Toll Roads and Free Roads* was issued, the Pennsylvania Turnpike had already destroyed the bureau's assumptions of traffic flow. The superhighway the BPR predicted would carry only 715 cars a day was carrying nearly ten thousand.

War put all highway plans back in the file drawers. Automobile production ceased. But the romantic examples of Futurama, of the Pennsylvania Turnpike, even of the Autobahns themselves fed into the larger dream of the postwar world that the war years nurtured. In 1944, with an eye to providing employment that would ease the projected postwar slump, Congress finally approved a very general plan for interregional highways based on the BPR scheme.

What could better sum up how easy life was going to be when the lights went on again than the vision of floating smoothly through a bright American landscape on a magic motorway? Superhighways, the highways of the future, were part of what Americans were fighting for.

PIONEER SUPERHIGHWAYS
The Pennsylvania Turnpike, 1940 (top), the first
constructed portion of what was to become the
Interstate System. (HERB GEHR, © 1940 TIME INC.)
An early German autobahn near Frankfurt
(bottom). (HEINRICH HOFFMAN, © TIME INC.)

5: Ike's Autobahns

For hours before the Pennsylvania Turnpike opened to the public on October 1, 1940, hundreds of enthusiastic drivers eager to try out the nation's first superhighway lined up at its entrance.

There were no speed limits on the four-lane divided highway, whose 160 miles ran from Harrisburg to Pittsburgh, and before the day was out automobiles would be streaking along the highway's great tangent, thirty-five nearly straight miles west of Bedford, at 90 miles an hour or more.

"America's Dream Road," the promoters and publicists tagged it. The driving was "magic," rhapsodized a *New York Times* reporter, who noted soaring above him "the hawks of the Kittatinny . . . their flight scarcely more swift or effortless than that of the driver of the new superroad."

It was the county's first full-fledged superhighway. Unlike the parkways, which banned trucks, it was to provide an important commercial route, shortening the truck trip between the steel mills of Pittsburgh and the port facilities of Philadelphia. To travelers returning from the World's Fair in New York, the Turnpike must have seemed a first realization of Norman Bel Geddes's visions.

For the first time, motorists encountered a highway where they could make full use of the modern car, which in two decades had tripled its power and speed. The motorist could average sixty miles an hour and drive the length in two-and-a-half hours, or about half the time the same trip took through the windy mountain strip of U.S. 30, the old Lincoln Highway. The lanes were a generous twelve feet wide, the shoulders capacious, and the sightlines excellent.

Pennsylvania was an appropriate place for the country's first superhighway. Not far away was the old Lancaster Turnpike, the finest wagon road the country had known, a rock road dressed in the manner of the best European roads, with gradually-sized stones fitted into one another in the Tresaguet method. The Conestoga wagon, with its high, boatlike ends to keep goods in on the grades and its wide wheels, had been born nearby; a historical marker indicated the first departure of such a wagon west to Ohio, the beginning of the westward migration.

The Turnpike crossed the Alleghenies, the nation's first barrier to western expansion, a patent demonstration of the country's need for transportation. During the French and Indian Wars, British General Braddock and his aide-de-camp George Washington had to blaze their own trail through the area on the expedition that ended in defeat at the Battle of the Wilderness. Braddock's road had become the forerunner of settlers moving west after the Revolution. The Whiskey Rebellion of 1794, an early demonstration of the need for good transportation, had flared up here.

It was also appropriate that the Turnpike made use of abandoned railroad tunnels to cross the Alleghenies. A casualty of competition among the railroads, the tunnels had lain empty, filling with debris and water, for years. The trucks the Turnpike (and the Interstates it would help inspire) were to carry would take much of the railroad's freight. On the Turnpike, trucks cut their transit time and fuel consumption to half of what they had been on the old Lincoln Highway.

Pennsylvania had supported many turnpike companies and the first railroads. Under Governor Gifford Pinchot, the noted forester, it had been a leader in creating farm-to-market roads in the twenties.

It already supported a major auto tourist industry that had dotted the Pennsylvania Dutch country with a series of windmill-shaped restaurants, half-timbered motels, and such attractions as "Noah's Ark" floating over the edge of the Lincoln Highway, offering a scenic vista.

In a few years the Turnpike would be hated by many drivers for its narrow medians, long, dull straights, and narrow tunnels, but for now it was the most modern auto highway the country had ever seen. Its opening came at a time when the country was on the threshold of the creation of its modern motor highway system. After World War II, the Turnpike would quickly be imitated in Maine, New York, West Virginia, Ohio, and New Jersey. Assigned the number I–76, it was the first portion built of what would become the Interstate Highway System.

President Franklin Roosevelt, caught up in the Presidential campaign, had considered an appearance at the opening. But Pennsylvania was a safe state for him and the role of his administration in the creation of the highway had been made abundantly clear. The Reconstruction Finance Corporation and Public Works Administration had financed it.

Studies for the "all-weather highway" to replace the steep and often icy U.S. Routes 11 and 30, the Lincoln Highway, began in 1934. The inspiration was the abandoned railroad tunnels which were to be adapted for highway use. Grades were to be kept to a maximum of four percent, as opposed to the eight or more found on the twisty, two-lane Route 30.

In the Depression, the idea of a major east-west turnpike for the state promised immediate economic stimulus—job creation—as well as the benefits of faster travel between Philadelphia and Pittsburgh.

The Turnpike established a model for the financing and administration of the first generation of superhighways that would be built, mostly in the Northeast, before the official start of the Interstate program. It was financed by a private authority, authorized to issue bonds by the state but without direct state funding. The bonds were to be retired by tolls from the finished pike. In this it resembled the old turnpike companies, like the Lancaster Pike.

In 1938 the Turnpike Authority attempted to sell $60 million of bonds on the private market without success. Then the Authority went to Jesse Jones of FDR's Reconstruction Finance Corporation and to Roosevelt-aide Harry Hopkins. With the 1937 recession threatening to drag the nation back into depression and the 1940 election looming, FDR pushed through a deal where the RFC would buy some $35 million of the bonds and the Public Works Administration would provide a direct grant of another $29 million to build the highway.

The terms of the deal required substantial completion of the highway by June of 1940—not just coincidentally in plenty of time for the election. Working at breakneck speed, crews completed the Turnpike in just twenty-three months, nominally meeting the June deadline. One hundred fifty contractors were involved, including a much-publicized one headed by a woman, Mrs. Margaret McNally.

Although its lines were far cruder than the curves of the German Autobahns, the Turnpike offered a similar system of limiting access with bridges and ramps. It even adopted the German system of modelling the restaurant and service buildings—whose concessions were let to Howard Johnson—on traditional regional architecture. They were sturdy, square stone structures like the big houses, barns, and inns of the colonial Pennsylvania landscape, foreshadowing the way the superhighways would replace small-time roadside eateries and motels with huge, franchised chains.

As war drew near, the Turnpike Authority emphasized the road's military utility. Maneuvers of mechanized National Guard units were held along the Turnpike almost immediately. The Turnpike, military planners noted, was also wide enough to serve as a landing strip for small aircraft. And the benefits to trucking the highway provided would make shipping war materiel east from Pittsburgh much easier.

The military uses of highways, the chief justification for roadbuilding in the old western territories, had become a concern again in World War I. At that time the railroads fell victim to a version of gridlock. The system virtually collapsed. Freight rotted on platforms. Lack of coordination between competing lines, facilities that had been allowed to deteriorate, and bad management caused either by complacency or by government regulation left vital war materiel languishing in depots. The government nationalized the railroads until the war's end, but they would never be quite the same again. 1916, the year in which the federal government first began aid to highways, was also the year in which the American railroad system achieved its widest extent: 256,000 miles.

The collapse of the railroads gave the trucking industry a chance to shine. Trucks for the front had previously been sent by rail from Detroit to the eastern ports. Now, they moved under their own power—and carried a load of supplies to boot—in huge convoys moving east from the Motor City.

By the end of the war, the long-term military uses of highways as an alternative to rail were on the mind of the newly returned General Pershing. Along the old National Road, Gallatin's track from the head of the Potomac to the banks of the Ohio, and hard by the future location of the Pennsylvania Turnpike, he dispatched in 1919 a cross-country convoy.

The motorized column of seventy-nine vehicles left Washington heading for San Francisco along the Lincoln Highway, with 260 enlisted men and 35 officers. Rolling through the small towns of the Midwest, sometimes passing impromptu crowds alerted to their arrival, the column of huge artillery tractors, tanker trucks, searchlight carriers, reconnaissance vehicles, ambulances, and motorcycles brought home to farmers and merchants the strange and distant modern war they had been hearing about. The soldiers stopped to visit and go fishing along the way, but more often their stops were occasioned by muddy roads and breakdowns. It took the column fifty-six days to make the trip—proof to the generals back in Washington of the inadequacy of the nation's roads for military or any other purposes. One of the officers on the expedition was young Dwight Eisenhower.

Twenty-five years later, as military head of occupied Germany, Eisenhower oversaw the "debriefing" of the Reich, the creation of a series of reports that included close study of the Autobahns.

The builders of the Pennsylvania Turnpike had also looked at the Autobahns. These were the first real system of superhighways. They were constructed according to principles of the modern auto road that had been worked out in theory well before: the division of traffic by a median, the separation of roads at intersections with ramps and bridges, the limitation of access to controlled points, and the shaping of the road's geometry for operation at a specific, constant, high speed. They were designed for safe operation at up to a hundred miles an hour. There were no speed limits, reflecting a German passion for cutting loose on the open highway.

Hitler himself loved to get out on the highway. Wrapped in a long leather greatcoat, traveling from a speech in one town to another, he would sometimes order his chauffeur to push the speed of his Mercedes to seventy or eighty miles an hour, passing confused drivers along the way. He particularly enjoyed passing American cars—imported Packards or Oldsmobiles—as a token of the superiority of the original German product.

The Autobahns were born of a mixture of economic, military, political, and aesthetic motives not terribly different from those that were to create the American Interstate system.

There exists a photograph of the first work crews heading off to begin work on the Autobahns in September 1933. It shows a neat social cross section of workers, many of them in country costume, others with the ragged appearance of the formerly unemployed, marching with shovels and picks over their shoulders, led by two uniformed men—Dr. Fritz Todt, head of the Autobahn program, and the local party boss.

Hitler built the Autobahns to provide employment and military arteries, but also to make an ideological statement that sketched an ideal future in the style of the German past. As portions of the Autobahn system reached completion in the mid-thirties, the Nazi propaganda ministry began inviting groups of foreigners in for tours. The keynote of these tours was a trip in one of the regime's huge Zeppelins, where visitors, comfortably seated in the wicker chairs of the passenger lounge, gazed out the windows as the pride of Germany floated majestically along above the track of the new roads, their sleek, concrete ribbons slicing through quiet, flowered meadows, dotted with half-timbered farm houses and barns, or through fairytale fir forests.

This was the way the Nazis wanted the world to see the Autobahns, and it was, to a much greater degree than critics would admit, the way they liked to see them themselves. Hitler envisioned the Autobahns with the eye of a failed painter and would-be architect. There was a longstanding German tradition that linked travel and natural beauty. It went back to the *Wandervogel* movement of the late nineteenth and early twentieth centuries and to the Romantics before them, and tied neatly into the mysticism of *Blut und Boden*.

The American embassy in Berlin returned regular reports on the progress of the Autobahns and pictures of them were reproduced in popular magazines. The reports turned in by the riders on the Zeppelins were filled with admiring images of the beauty and virility of the system. A British professor, R.G.H. Clements, who rode the Graf Zeppelin as part of a General Roads Delegation reflecting, he wrote, "many shades of opinion," came back with phrases like "sheer clean beauty" and "vigorous sweeping curves." Propaganda minister Goebbels must have been happy.

F.A. Gutheim, writing in the *American Magazine of Art* in 1936, tells of two roomy concrete lanes in each direction, separated by a grassy median. "To either side considerable strips of land of varying width are preserved as integral parts of the scheme, preventing access to the roadway and possessing definite aesthetic advantages."

Their design was highly scientific. "All curves are wide and well saucered for high speeds; on the Frankfurt-Darmstadt stretch I rounded a curve at over sixty m.p.h. without touching the steering wheel, the 'set' of the car's gravity in the bank carrying it around and straightening out the wheels automatically with the road." The writer praised the "simplicity and crispness of design" of the bridges and other features.

Exits and entrances occurred only every twenty miles or so, in keeping with the program of reserving the highways for high-speed express traffic between principal cities. Villages were bypassed. Some 2,326 miles of the system were finished when the war brought construction to an end in 1942.

Germany, birthplace of the automobile, had a long tradition of experimentation with modern highways. A private organization called AVUS had built a modern divided highway as a demonstration as early as 1913; completion was delayed until 1919. It ran for just six miles through the Grunewald park but served as a model for later designs. During the flourishing of the Weimar Republic, discussion of a modern freeway system was a constant topic. Private organizations pushed for the right to build modern highways, using a toll system, but the government did not pass the necessary legislation. The German magazine *Die Strasse* was influential in spreading the gospel of the superhighway. Various Autobahn proposals were made, with no result, during the Weimar Republic.

In 1930, with these projects in the background and millions out of work in Germany, the engineer Fritz Todt gained Hitler's attention with his Autobahn plan in a paper called "Proposals and Financial Plans for the Employment of One Million Men." The title indicated the program's emphasis on economic stimulus. When Hitler came to power in 1933 he almost immediately named Todt to head an Autobahn program.

Fritz Todt—whose name lends itself to a pun on the German "Tod"—"death"—was the quintessential Nazi technocrat. During the Weimar years, he rose to become manager of the Munich construction firm of Sager and Woerner. He joined the Nazi party early—in 1922—and became one of Hitler's closest associates.

After the Nazis took power, he created the dread Todt Organization, which was to use slave labor in building factories and fortifications. He fortified the West Wall—the Nazi equivalent of the Maginot Line—and the French coast and directed the building of railroads and bridges on the eastern front.

The Autobahns, of course, had tremendous military utility to a regime whose military adventures were based on mechanized assaults and the *Blitzkrieg*. But the other motives for their construction were also important—easing unemployment and boosting the economy. That this involved direct expropriation of land, the diversion of unemployment insurance funds into highway bonds, and the use of forced—if not slave—labor bothered no one. And they were excellent propaganda, both to the outside world—the foreign visitors who toured them by Zeppelin—and, despite the fact that only the elite owned automobiles, to the German public as well. It was true that the traffic represented chiefly the powerful and well-off, but the implied promise of the Autobahns was that, under Hitler, people would soon be driving their own cars—cars like Dr. Ferdinand Porsche's prototype Volkswagens.

At least as important, however, was the Nazi notion that building the Autobahns was an exercise of German "strength and will"—a flexing of economic and administrative muscle, what one Nazi official called part of the national "return to health," a "test mobilization," a rehearsal (as it has come to seem in retrospect) for war, in which the regime's half-baked ideas of national character and spirit mingled with its aesthetics.

Similar motives—economics, prestige—were to inspire the American superhighways. The irony is that the Interstate system was itself inaccurately sold to the public in the mid-fifties as a "system of national defense highways" to facilitate movement of men and materiel and, illusion as cruel as any of the Autobahn's, to evacuate cities in the face of imminent nuclear attack.

It was with these experiences in his background—the lessons of the military uses for highways, the model of the Autobahns—that Eisenhower approached the building of the Interstate system.

The Interstate program was the last New Deal Program and the first space program, combining the economic and social ambitions of the former with the technological and organizational virtuosity, the sense of national prestige and achievement, of the latter. The planners of the system expanded the public works aspect of highway building into a vision of Keynesian pump-priming and economic "fine-tuning," directed under Eisenhower through private industry. They linked the economic vision to a dubious one of national defense. They adapted the aesthetic—and implicitly the social—values of the American parkway tradition to the new suburbia. They looked back to the German Autobahns for standards of beautiful and efficient engineering. And they answered a public demand for the realization of the utopian, technological future outlined by the streamlining visionaries of World's Fairs and Sunday supplements.

Although it was almost certainly the major domestic achievement of his administration, Eisenhower barely mentions the Interstates in his memoirs. This may be because the system as realized had little to do with his original proposals. Not that the highways looked different or were located any differently—the map of 1944 and the design standards of the postwar years were realized in the final system. What changed was the means of financing and adminis-tration of the program.

For a decade after the end of World War II, there was public awareness of a road crisis, "a highway situation." The arsenal of democracy, shifted to peacetime production, turned out automo-biles with the same speed it had produced tanks and planes, and the postwar boom sent more and more people onto the road. But the political wrangling delayed the beginning of the highways mapped out in the 1944 interregional highway plans.

Toll roads on the model of the Pennsylvania Turnpike flourished in the Northeast, but no grand national plan existed to link them. Then, to get things moving, Eisenhower in 1954 appointed Lucius Clay to head a committee designed to solve "the road problem" and finally get the Interstates going. The appointment was an immediate indication both of Eisenhower's vision of the Interstates as a weapon in the Cold War and of their political ancestry.

Clay was the scion of a political family—his father was a United States senator, and Henry Clay was his great-granduncle—a mil-itary engineer trained at West Point, a tough and efficient admin-istrator with a remarkable memory for detail.

In 1948, the Soviet forces occupying East Germany blocked off the Autobahns to West Berlin in an effort to make the Western allies release their hold on the former German capital. The response was the famous Berlin Airlift, and the man who ran that airlift was Lucius Clay.

The airlift made Clay famous. He returned to the United States in 1952, and was instrumental in working behind the scenes to assure Eisenhower the Republican nomination for President in that year. By 1954 he had become head of the Continental Can Company.

The plan Clay's committee came up with was quite reasonable: creation of a national highway authority to build the Interstates using federal bonds. It reflected Eisenhower's basic philosophy, although Eisenhower at first hoped the roads could be paid for by tolls.

Eisenhower saw the Interstate as a demonstration of the "right way" to manage the economy, an appropriately Republican version of a Keynesian program. It avoided Roosevelt's "excess condemnation" idea, which came close to expropriation. (The Clay Committee did consider condemnation, but, aside from the major ideological objections, there were administrative problems as well: drawing the line and handling the sales presented opportunities for huge corruption.)

Like Roosevelt in the late thirties, Eisenhower and his economic adviser Arthur Burns saw the Interstate system as a tool to manage the economy. But Eisenhower believed in priming the economic pump with money funneled through the private sector, not in makework programs like the WPA. If that meant helping big business, so be it; the automobile, directly and indirectly, accounted for nearly a fifth of the whole economy. Building roads would not add to the efficiency and productivity of that economy, but would aid its largest components directly: the auto makers, of course, and steel and rubber and plastics and construction. This was a view of internal improvements in the tradition of the Whigs and of Henry Clay's American System.

Eisenhower's views on the matter were not as simple as those credited to his Defense Secretary, Charles Wilson, formerly of General Motors, to whom was credited the famous line "What's good for General Motors is good for the country." But they were accurately summed up in what Wilson *really* said: "I've always believed that what was good for the country was good for General Motors, and vice versa."

In addition to the economic impact of its construction, Interstate boosters described the beneficial effects that the highways would have on the ease of operation of the whole economy. *Things would move better.* Again and again in the arguments in favor of more highways, proponents talked about how a highway "turned space into time." With a fast road you no longer talked about how far you had to go, but about how long it would take you. For commuters, truckers, business travelers, or tourists, the Interstates could extend the isochrones (lines delimiting the distance one could travel in a given time). If they did not create new places to go, what they did was equivalent in effect; they made it possible to go more places in the same time.

Eisenhower, too, believed that highways would bring benefits for the whole country. In speaking of the way better highways would provide "greater convenience, greater happiness, and greater standards of living," he echoed widespread public opinion. And there was a clear vision of what sort of living the highways implied—a vision of the American Way of Life that was a sum of clichés about the Eisenhower fifties, a decade at whose end the suburban population would equal that of the cities themselves.

When Nikita Khrushchev was planning to visit the United States in the late fifties, Eisenhower knew just what he wanted to show him: Levittown, a powerful demonstration, in Ike's mind, of the success of the American system in giving every worker a house as efficiently mass-produced as the car he drove. (Khrushchev, for his part, wanted to see Disneyland.) Part of this success, naturally, involved highways to get him to and from work. The new highways would shorten the time it took to get to the suburbs, effectively expanding the space of the city, trading off space and time in the traditional American manner.

Popular opinion supported this vision, but it was big business that led in mobilizing it. In 1955, the Ford Motor Company published a book called *Freedom of the American Road* boosting road-building. It was called "an action book of things you can do" and included articles by highway boosters from Thomas McDonald to Robert Moses, showing the benefits of good roads as well as case studies of how "enterprising civic-minded citizens" had gotten new highways built in their localities. The message was that good roads were good Americanism. The appeal was patriotic, boosterish, almost folksy. Henry Ford II wrote in the Foreword: "We Americans always have liked plenty of elbow room—freedom to come and go as we please in this big country of ours." Lobbying for good roads, Ford wrote, was "democracy in action," and typically American "resourcefulness."

The "National Defense" tag added to the highway program reflected Eisenhower's background. He came from a new military tradition: the administrative one. He had attended not only the Army War College, but the Army Industrial College. As an officer under MacArthur and Marshall he had become the ultimate aide-de-camp: the soldier as technocrat. His was not only the military that created the logistics miracle of D-Day, but the military that built the Pentagon (a structure served, incidentally, by one of the most modern systems of feeder highways in the world), the military of advance planning and of crash programs, like the Manhattan Project. This combination of the industrial ethic with the military one was to shape the entire postwar economy. Not by accident was Eisenhower, discouraged at the end of his terms by his struggles with the Pentagon bureaucracy, the man who popularized the term "military-industrial complex." But, like the space program, the Interstate program Eisenhower had promoted was a product of just this complex.

The Clay Committee represented the military way of solving "the highway problem" that the postwar boom had made more acute: the way of the "project," "the operation," the "task force." You appointed a task force, worked out a plan of operations, and then launched the project.

But that was not the way things worked in Washington, particularly when it came to highways. Highway building had become too big a pork barrel, with too many constituencies, to allow for centralized planning. Highway policy pitted the northeast against the west, rural against urban, bond men against tax men, Keynesians against fiscal conservatives, planners against engineers, federal against state bureaucracies. Highways provided a virtual model of the conflicts not only of government but of the whole society.

The two most important highway constituencies were AASHO, which had become an interests group far more powerful than any single government agency, and the rural roads interests that the BPR had done so much to develop over the years, and which continued to see highways in twenties terms—farm to market, rural development. These groups were particularly well represented in Congress by conservative representatives with long seniority, men like George Fallon of Maryland and John "Big Klu" Kluczyncki of Illinois, a former truck driver.

A survey of the members of the Clay Committee suggests the other powerful interests involved. In addition to general interests—the U.S. Chamber of Commerce and the American Automobile Association—and rural interests—the American Farm Bureau Federation, the National Grange, the National Association of County Officials—the truckers were well represented. Testimony was heard from the Truck-Trailer Manufacturers Association, the Independent Advisory Committee to the Trucking Industry (including David Beck of the Teamsters), the Private Truck Council of America, and the American Trucking association.

The war had made trucking interests even stronger, and their resistance to higher taxes was a serious obstacle to building the Interstates. They finally came out big winners from the system: the efficiencies the superhighways added to their business were nearly twice as much per mile as the increases in taxes they faced.

Industry interests were well organized and numerous. They included such organizations as the Portland Cement Association, the Asphalt Institute, the Automobile Manufacturers Association, the American Institute of Steel Construction, the Highway Users Federation (despite its name, an industry front), and the American Automobile Association.

New interest groups had arrived as well, particularly the urban planners who saw expressways as a tool for rebuilding the cities. In the late forties, it was believed that development of the urban fringe "choked" the central city. Terms like "blight" and "rot" were becoming popular. If the urban circulatory system failed, the implication was, then something like gangrene would set in. Thus groups like the U.S. Conference of Mayors (whose spokesman was Robert Moses), the American Municipal Association (later the National League of Cities), and the National Association of County Officials also lobbied for superhighways.

The established roadbuilders had most to fear from any change in the control of the highway purse. The Toll Roads Association turned down an offer to testify before the Clay Committee. The engineers of AASHO feared being dictated to by a national authority, and BPR continued to fear toll roads, which, when the federal role had declined in the forties and fifties, had boomed in the Northeast.

Lurking behind all the specific conflicts was a larger conflict, the traditional argument over federal power versus state power. It took the form of a conflict between "pay as you go" and long-term bond finance—raising issues of national debt and centralized monetary power that went back to Alexander Hamilton and Thomas Jefferson and the controversy over the Bank of the United States.

The federal role in highway building had slim constitutional supports to begin with. The first federal aid programs, dating from 1916, had begun under the guise of improving postal service. Conservatives in 1956 still resented the federal role; they pushed for a "pay as you go" scheme with strong state control. "It was a question," said Frank Turner, the secretary of the Clay Committee, "of a national system of highways versus a system of national highways."

What emerged from a Congress wracked by the contentions of these groups was the creation of a Highway Trust fund, financed from federal taxes on gasoline, tires, and other automotive products. This fund would assume the cost of ninety percent of Interstate construction, with the remainder and all maintenance left to the states. The highways would be constructed to federal standards, along routes approved by the states. The original Clay Committee idea of bond financing was rejected, due to the efforts of conservative congressmen and senators, in favor of a "pay as you go" plan in which each year's construction would be financed by that year's projected highway tax revenues.

The toll interests were protected. Toll roads could be retroactively financed with Interstate money. Eventually, the tolls were supposed to be removed, but they rarely were. With federal funds added to the toll receipts, some of the toll agencies, like the New Jersey Turnpike Commission and the New York-New Jersey Port Authority, became virtual rogue agencies, subject to little political control as they built more and more highways.

The Highway Trust Fund became a financial juggernaut on which all the various interest groups had comfortable seats. It quickly assumed the same political sanctity enjoyed by Social Security. It essentially removed control of highway spending from Congress and the White House. Studies consistently showed that highway construction had a very low multiplier effect—the dollars spent mostly stayed in the construction industry. As finally structured, far from working as the macroeconomic governor or pump-primer Eisenhower and others—going back at least as far as Roosevelt— had originally envisioned, it had exactly the opposite effect. The Fund System guaranteed that added miles of travel produced by good economic times would produce more taxes and therefore more roads, putting inflationary pressures on the economy. If it had any macroeconomic effect, it was to accentuate the fluctuations of the economy. This was acknowledged implicitly by Congress in 1958, when, with the country in recession, it temporarily suspended the pay-as-you-go provisions of the plan to create construction for immediate economic stimulus.

Eisenhower was proud of the Interstate legislation. The federal highway administrators soon began to issue statistics illustrating just how dramatic an achievement the program was. It was to be the greatest public works program in history. It would move enough earth to cover the state of Connecticut knee-deep; it would claim enough land in rights of way to equal the acreage of Delaware; it would pave a surface equivalent to that of West Virginia. The concrete it consumed would build six sidewalks to the moon, and it would use as much lumber as produced by 400 square miles of forest, as much drainpipe as the water and sewer systems of six Chicagos.

The Interstates were only a tiny part of the whole national road system, an average of perhaps twenty percent of road spending in the twenty years after the program began, and only two percent of the total mileage, but they became the model for new highways. The states strove to match their standards. The states had to pay for only ten percent of the Interstate costs, but nearly half of the primary federal aid system, all of their own system costs, and all maintenance costs. In many states, highway building in the 1960s accounted for well over half of all public capital spending.

While the "National Defense Highway" tag the Eisenhower administration added to the Interstate title was a convenient one in the fearful years of the mid-fifties, it was also a real part of the scheme. With the threat of atomic warfare looming—the Soviets had exploded their first atomic bomb in 1949—it seemed provident to provide evacuation plans for the cities, even if, privately, officials knew there was no way to accomplish this without advance warning. Eisenhower, after participating in a test of the military readiness system, declared that we should get to work at once on better highways to evacuate the cities.

It was assumed, too, that in case of war, either nuclear or conventional, the centers of cities would be destroyed by bombing as they had been in Europe. The result was a bizarre combination of incentives—highways to foster the happy, familial, baby-boom suburbs and highways to assist in a potential cataclysmic conflict. And yet somehow they fitted together: the hula hoop was a refuge from the mushroom cloud.

But what the Interstates—like the space program—shared most directly with the military was a system of thought and organization. The Interstates were the apotheosis of that great American love of the project, the program, the system, the module.

Like many such programs, the Interstates took on a bureaucratic and political momentum that was almost impossible to resist. While other transportation needs were neglected, highways, with their unique claim on tax funds, prospered. Even when the Fund was revamped in the 1970s, only token provisions were made for shifting some of its revenues to railroads, bus lanes, and other forms of transportation. The gas tax was regarded as money that belonged, by natural right, to highways. "It would be immoral to shift that money," said highway administrator Frank Turner at the time. As sacrosanct as Social Security, the Highway Trust Fund had come to enjoy the status of a natural, constitutional right for highway builders.

Not incidentally, the Fund also created a negative energy policy. Taxes were raised per gallon, not as a percentage of sales price. Thus highway builders had an incentive to increase gallons consumed. When the gas crunches of the seventies raised prices and cut gasoline consumption, they also reduced the amount of money available for construction and maintenance. The infrastructure crisis of the late seventies was a direct result of the taxing policies of the mid-fifties.

The Interstate system was also a program that was by nature open-ended, one that many people doubted would ever be finished. Politics inevitably added miles to the length of the system, and increasing traffic would require more roads to keep it moving. When a Democratic administration came to power in 1961, it promptly raised the gasoline tax from three to four cents to finance more Interstates, most of them in urban areas where Democratic support was stronger and the howls of motorists caught in traffic jams were louder.

Even more significantly, the distribution of funds, after the first few years of the program, was adjusted to pay states not according to their population but to their proportion of the total system. This provision helped states like California, Florida, and Texas, less populated states with long stretches of planned highway in what would become known as the Sunbelt.

The vision of the highway as economic driver included the even development of growth, throughout the country. The Interstates and their relatives did more than any single factor, save perhaps the location of the bulk military facilities and contractors, to produce the boom of the Sunbelt.

The belief that roads would bring prosperity was equally held by Republicans and Democrats. Appalachia, for instance, had been almost completely avoided by the Interstate program because it contained few of the major population centers the system was supposed to link. John Kennedy planned to expand the highway program to include the area, and Lyndon Johnson, realizing these programs and looking back to his mentor, FDR, made roadbuilding the centerpiece of his War on Poverty. Of the five billion dollars spent on the Appalachian development program, three billion went for highways. It was difficult to build superhighways through the mountains, and one stretch of U.S. 64 through western North Carolina literally kept slipping off the mountainside. It brought several contractors to grief before it was finally completed. The highways were supposed to bring in industry and jobs, but in fact they brought in mostly tourists.

Above all, the Interstates were aimed at "keeping America moving." To the engineers, this meant no more than keeping traffic moving; to the interest groups, it meant keeping the flow of their products moving; to the politicans, it had economic, military, patriotic, and almost spiritual implications. The intent of the Interstates was less to bind the country together than to realize an almost mystic principle: to keep things circulating, flowing, to keep the road open.

In 1978, when the last link of I-80 between Boston and Seattle was completed, a ceremony was held near Blue Lake, Minnesota. The final stretch of highway was marked with gold-painted pavement, an echo of the Golden Spike that completed the Transcontinental Railroad. But the engineers had missed the point. The Interstates were created not so much to bind the nation together as to keep things flowing.

A more apposite ceremony was held in 1984, near the small town of Caldwell, Idaho. On that day, with officials and reporters looking on, "red-eyed Pete," "the last stoplight on the Interstates," was removed, placed in a coffin, and ceremonially buried.

The urban highway as megastructure. The proposed Lower Manhattan Expressway, a transportation core for urban renewal. As envisioned in the late sixties by architect Paul Rudolph, this project would have sliced across Canal Street and through SoHo, Little Italy, and Chinatown. (COURTESY PAUL RUDOLPH)

6: The Superhighway in the City

"One thing you can set down as sure is that cities are doomed."

—Henry Ford

In Le Corbusier's 1922 sketches for his "Ville Contemporaine," elevated superhighways soared among skyscrapers, over open parks and past terrace cafés where patrons happily sipped espresso beneath delicately-leaved saplings and watched the steady stream of traffic.

This was the original modernist vision of the twentieth century city: the American gridiron pattern of streets turned over to pedestrians and, superimposed on it, a radial system of superhighways carrying auto traffic. Surrounding the city there was to be a greenbelt of parks and a circumferential superhighway.

It looked lovely in the sketches, all green and open. It didn't, in fact, look much like a city at all. The sketches could not hide the fact that most of the people in love with the idea of urban superhighways did not really like cities very much.

Almost all the early supporters of superhighways believed either that cities would eventually wither away as society, helped by decentralizing highways, evolved toward a pastoral, suburban ideal, or that the cities would be replaced by new cities of a very different pattern.

In adapting his Ville Contemporaine plans to a proposal for re-creating Paris, Le Corbusier did not hesitate to call for a demolition of the heart of the city that would have equaled anything accomplished by bombers. Le Corbusier was good at cities built on tabula rasa—in fact his thinking demanded erasing the tabula.

Frank Lloyd Wright, who claimed to disagree with most of Le Corbusier's ideas, nonetheless turned out urban plans that were similar to his. Wright wrote a book called *The Disappearing City*, and shared the conviction that automobility would mean the eventual end of cities. Wright's Broadacre City was a contradictory vision, a "decentralized city" where superhighways blended with individual one-acre estates in that utopia he called "Usonia." Throughout his life, Wright spoke almost casually of the demise of cities. It seemed to him as inevitable as the end of the horse and buggy, and the modern highway systems he watched being built were to be the agents of the city's destruction.

One day in the summer of 1959, Dwight Eisenhower was being driven from the White House to the presidential retreat at Camp David, where he could escape the hot weather in Washington. On the way, he happened to pass a construction site for an Interstate leading into Washington. As soon as he reached Camp David, the story goes, Eisenhower telephoned highway officials to find out what was going on: it was his impression that the Interstates were not to run into the cities.

The highway officials, both then and again and again in later years, expressed astonishment: hadn't Eisenhower seen and approved the maps, with the urban routes plainly marked? Could he have forgotten how the support of mayors and congressmen from big cities was enlisted for the program? Didn't he know that the objective of the whole project had been defined as linking most American cities of over 50,000 population? Where, after all, did he think the majority of Americans lived? And how did he expect to achieve his treasured goal of using the Interstates to evacuate the cities in the event of war?

It was either startlingly direct confirmation of the image of Eisenhower as a president out of touch, a distracted figurehead floating in blissful ignorance above the nitty gritty of policy, or evidence of a willful refusal to see what he did not want to see. Yet whatever the source, Eisenhower's refusal to consider the consequences of Interstates in the cities was shared by much of the public.

Perhaps the best explanation for Eisenhower's behavior is that the vision of cities crisscrossed by superhighways, however parklike, simply did not accord with the vision of America nourished by a man who grew up in Texas and Kansas, who had spent nearly a decade out of the country, much of it contemplating the ruins of urban Europe, and who, like many Americans, preferred spending time in a place like Camp David to spending it in Washington or any other city. Eisenhower was proud of a country where every working man could drive his car to his home in the suburbs.

The Interstates had always, necessarily, been planned to include urban links. It was traffic congestion that provided the greatest impetus to their creation, and the congestion had always, from its beginnings in the 1920s, been in the cities. Superhighways were supposed to improve urban life. The 1944 report that sketched out the beginnings of the Interstate program poetically envisioned them as "elongated parks bringing to the inner city a welcome addition of beauty, grace, and green open space."

The Interstates had been laid out around as well as through the cities. Military thinking helped dictate this policy. It was assumed that World War III would destroy urban cores as World War II had destroyed them in Europe, and routes around them would be necessary to keep military traffic flowing. Now it seemed almost as if Eisenhower was writing off the city centers in advance of the war that would destroy them.

Other highway visionaries saw urban freeways chiefly as a way to get out of the city. In Futurama and in his book *Magic Motorways*, Norman Bel Geddes was obsessed with a suburban vision that differed from Wright's only in expecting continued commuting to city centers. And the odologist cannot help but suspect that Bel

Geddes only kept the central cities in his plans at all because of the opportunity they gave him to design lovely skyscraper models.

When highways go into the cities, he charged, they "cause congestion and confusion." The job of the superhighway was to connect cities, he argued, and it "would only bungle the job if it got tangled up with a city." It would, he said in a revealing phrase, "lose its integrity."

"A great motorway has no business cutting a wide swath right through a town or city and destroying the values there; its place is in the country, where there is ample room for it and where its landscaping is designed to harmonize with the land around it."

Bel Geddes's alternative was to consider highways as routes "between the environs of cities." He skimmed over the problem of relating his magic motorways to existing urban cores, admitting that there would have to be "feeder roads" from the great loops around the cities into its core, but suggesting that they should not be very large.

But this plan did not jibe with his other goal—expanding the commuting radius. A city is a "*working* entity," he said, citing the thinking of utopian Robert Owen, while the country is a "*living* entity." In truth, he wanted to strengthen the suburbs, reduce the importance of cities, shift population into remote corners of the country, and reshape the economic map with his superhighways.

"If cars could go twice as fast as they do today," he wrote in *Magic Motorways*, "the accessibility of the city's surrounding area would increase fourfold." In the language of the classic suburban vision, he added that "this would make possible a general thinking out which would give everyone freedom."

Bel Geddes provided a map of Los Angeles comparing the forty-five minute isochrones—the distance one could travel in that time at the two different rates of speed—with the caption, "A modern highway system would extend a city's commuting radius six times."

Such an extension would have choked Bel Geddes's feeder roads with traffic. It would necessarily have increased congestion. Expanding the commuting radius by six times would draw into the city not six times as many people but nearly thirty-six times as many—the expansion was an expansion of area, not just distance.

This was a mistake that many other planners, forgetting how radial highways serve not just their borders but whole areas, were to make.

Bel Geddes's ideas worked best for cities in the middle of plains, with easy expansion in all directions. He did not want to have to deal with such knotty problems as those presented by older cities, like Boston or New York, built on water transportation, where ocean front prevented the creation of beltways, rivers created chokepoints to land transportation at bridges and tunnels, and density and economics made universal automobile ownership impossible. His ideal city was modelled on St. Louis, a middle-American city that was easy enough to reimagine.

More important was the fact that many of the men who built freeways in the cities harbored an overt or latent hostility to them. Robert Moses, who called Bel Geddes's ideas "bunk," conceived his early parkways and then expressways on the premise of getting the middle class out of the city and into the state parks he built. He clung to the guiding notion of the early parkways as "pleasure roads," from the days when driving was still conceived as a recreational activity, like strolling or horseback riding. Later this vision became one of getting out to suburbs located near those parks—to Ronkonkama or Rye. All of Moses's highways, the parkways and the expressways, were to be linked to one another. In the process, they ended up destroying large portions of the city and cutting others apart.

The ideals of escaping to suburbia and the countryside that Moses and Bel Geddes shared, ironically, dictated that highways would run into the cities—if only to get people out of them. Bel Geddes was right about one feature of the world of 1960: in that year, suburban population exceeded urban population.

But for all the arguments against building superhighways in the cities, other planners were touting them as means of urban redevelopment—a way to keep the city alive. Familiar as confrontations between expressway builders and opposition coalitions have become, it is easy to forget that until the end of the sixties expressways were a standard part of the liberal program for urban renewal.

"Urban decay," an issue as early as the thirties, was described for many years in circulatory terms. The central cities, it was thought, were dying because people were moving out to the suburbs, and they were moving to the suburbs because the new highways made it easy to do so. The problem was therefore a circulatory one; you simply had to bring people back into the center cities.

Even as Eisenhower was making his indignant phone call, city officials all over the country were figuring out ways to designate as Interstates freeways they had long dreamed of, thus handing over to the federal government ninety percent of the cost of building them. The Interstate program was to provide federal funding for most of these schemes. Over the years, Interstate funding has become, in effect, the single largest federal urban aid program.

Eventually, the three percent of Interstate miles in the cities and the other urban expressways were to become something like a highway Vietnam. Highway planners began to advance arguments that sounded like destroying the city to save it. Thus they justified the destruction of millions of units of housing—housing that in the inflated and debased construction industry of the postwar years was literally irreplaceable—and of small businesses that frequently ceased operation rather than move, the displacement of hundreds of thousands of people from their homes—the slums would have had to be torn down any way, the engineers always said—and the invasion of previously sacrosanct parks. In response, for the first time in American history there emerged a wide body of public opinion opposed to highway building. It began in the cities, but soon it was everywhere.

The highway builders couldn't understand it. How could the public not appreciate all the engineers had given it? Some sounded hurt, misunderstood. "We were environmentalists from way back when," said Frank Turner, after he had retired as director of the Federal Highway Administration. Others were resentful; one administrator termed the opposition "terrestrial anti-vivisectionists." And Alfred Johnson, head of AASHO in the early seventies, took a the-hell-with-them attitude, wondering whether it wouldn't have been better to deny Interstates to the cities in the first place.

• • •

It was clear in advance to many observers just how dramatic the impact of superhighways on the city would be. In 1957 Lewis Mumford predicted quite accurately the effects of the Interstates which he knew very well would run through the hearts of the cities. Mumford argued that "the highway engineers have no excuse for invading the city with their regional and transcontinental trunk systems: the change from the major artery to the local artery can now be achieved without breaking the bulk of goods or replacing the vehicle; that is precisely the advantage of the motorcar."

Mumford compared the freeways, most of which were necessarily elevated above the existing street network, to the much criticized elevated rail lines, with their noise, the ugly pillars and the shadows that turned the blocks around them into a mottled jungle floor.

Highway building in the city followed a pattern traced by power lines, which, first installed at street level, were soon elevated on poles, and then, in congested areas, put underground. Railroads followed the same pattern, with the noisy, sooty "el," raised above street traffic, finally placed beneath it. As it became clear that simply widening ground-level streets would not solve traffic problems, the expressways were lifted over the traffic grid on piles. Beginning in the fifties, it became clear that in many cases the expressways would have to go underground too.

In each case, this process happened most rapidly in the most crowded areas, where the effects of these constrictions on the human environment, whether in the form of pollution, noise, or ugliness, created sufficient pressure to spend the extra money necessary to bury the offending systems.

Urban superhighways fell into several patterns of dispersal. One was the old-fashioned focusing of several radii on a central core, as in Indianapolis, where a record seven Interstates converged. Another was the laying out of freeways as a flexible grid, a network over the whole urban area, as in Los Angeles. Most successful in younger cities was the modified spoke and wheel approach, where the wheel of the beltway or circumferential highway became the city's organizing growth ring.

The choice of pattern and the resulting side effects of freeways in the cities corresponded almost exactly to the age of the city. In port cities originally built on a system of water transportation and with high population density, they could hardly intrude without wholesale destruction. Choices for their location were limited by the chokepoints of bridges and tunnels. In the river cities of the Midwest—St. Louis, Memphis, Cincinnati, and so on—they tended to separate the city from its riverfront. And everywhere they were built their appetite for land—land that, not incidentally, would be removed from the property tax rolls—dictated that it would be the poorest communities—ethnic communities, black and Hispanic communities—which were destroyed to make way for their passage.

In Boston, the nightmare Downtown Artery (the name summed up the clichés of the circulatory theory of urban renewal) which sliced off the city's Italian North End and much of the harbor area, was an example of this; or the mad attempt to run a superhighway through the old French Quarter of New Orleans. In Philadelphia, the only place the planners could find to run the Schuylkill Expressway—quickly nicknamed the "Surekill"—was through Fairmont Park. San Francisco in 1959 banned all future freeway construction, leaving the Embarcadero Freeway, one of those disastrous downtown walls, an unfinished hulk that terminated in a huge leap as attractive to potential suicides as the Golden Gate Bridge.

In cities with growth and transportation patterns already widely dispersed—like Los Angeles, thanks to its farflung rail network—the oversized grid of the freeways fit into the local grid with less conflict. In cities just undergoing their major growth in the fifties and sixties, cities with plenty of open space remaining—in Atlanta, Dallas, and Houston—they were most successful. Fewer established neighborhoods lay in the path of the highways, and growth developed along the beltways and arteries they created.

As opposition to freeway construction grew in popular support even among the most avid drivers, the usual American search for villains began. It was almost like the railroad days again. Critics condemned "the Road Gang," the automobile, petroleum, and tire companies, for building highways we didn't need and crippling efforts to build and sustain mass transit.

By 1972 there was even a full-fledged conspiracy theory. The automobile companies, which had shown themselves at their bull-headed worst in resisting efforts at improving the safety standards of their products, were the natural villains to choose. General Motors, as the largest of these, the pioneer of the annual body change and planned obsolescence, was seen as the ringleader of the conspiracy.

In the thirties, a General Motors holding company called National City Lines had begun buying local urban rail lines in cities from Nashville, Tennessee, to Butte, Wyoming, and converting them to bus service. Later, oil and rubber companies became investors in this plan. Los Angeles and San Francisco trolley systems were shut down by the group; New York trolley lines were replaced almost overnight by buses. Critics have charged that this conspiracy—some of whose principals were convicted of Sherman Anti-Trust violations in 1949—was the reason Americans began the switch from streetcars and trolleys to private auto and freeways.

It is not so simple. The power of the auto oligopoly surely sped up the process, but it was already the determination of the American people to put a car in every garage that had made those companies so powerful. Popular preference for the private car, inexpensive gasoline, and highway expansion killed the short rail lines, aided by the bad management that had long characterized American railroads. The appeals of the car and the road made the public easy marks for exploitation by big business.

The truth is that virtually everyone who can afford an automobile in this country prefers using it to mass transit. So do many who can't afford an automobile, as the strength of the used car and repossession trades attests. Americans hate waiting on rail platforms far more than they hate waiting in their cars in traffic jams. They talk of being "herded" onto rail cars and the verb "to be railroaded" is a fixed metaphor of our language.

Of equal importance to the automobile's ability to carry its driver where he wants is its ability to do it *when* he wants—and the fact that he rides privately, enclosed in his own space. Sociologist Edward Hall theorized that different nations and cultures possess their own "space bubbles," their implicit sense of how close individuals are permitted to get to one another. Perhaps it is the nature of the American space bubble, the demand for private space in even the closest of social juxtapositions, that explains our resistance to mass transit.

In the sixties and seventies, certain forms of rail transit became an upper-class liberal cause, supported by center-city business. But mass transit was doomed by its very name: Americans don't like to be thought of as masses. Even stuck in the midst of a traffic jam, the driver sees himself as a free, democratic atom.

Subway systems had become hugely expensive by the 1970s. San Francisco's Bay Area Rapid Transit ran more than four times ahead of its original cost estimates, and eventual ridership was half what had been projected. It was a commuter's Cadillac, created to serve central-city enterprise and well-off commuters, and financed by the most regressive of taxes, the sales tax.

In many ways, BART-style mass transit plans repeated the mistakes of the big city freeways. They were capital-intensive and inflexible. While bus lanes, jitneys, and policies to encourage car pooling and for-pay ride sharing would utilize the existing freeways to produce far cheaper efficiencies, they were less politically impressive, offering no ribbons to cut.

Heavy rail systems are also very energy-inefficient. The energy used in their construction is equivalent to years of the operating energy of automobile or bus use. And even for regular operation, buses are between two and three times more energy-efficient per passenger mile than rail.

Had rail corridors been reserved along the right of way of highways, as was occasionally done in Detroit and Chicago, and proposed for New York, the costs of building systems would have been reduced. In addition, the attractiveness of rail would have been emphasized in the most direct way possible: the train speeding by the motorist stalled in traffic would have appealed to the same instincts for speed and convenience that put him in his car in the first place. It was the only chance, perhaps, that such systems ever had.

The next best thing to that sort of system is the creation of bus lanes—a relatively inexpensive, flexible, and neglected transit solution. Tax and other incentives to foster jitney services and private car pools have never been explored. Nor has much effort been made in major cities to stagger commuting times. The ineconomy of building any transportation system to meet a one- or two-hour peak of traffic is obvious, and many estimates hold that in most U.S. cities simply increasing the average occupancy of automobiles by a few tenths of a person—by buses, jitneys, or car pool incentives—would provide a reduction in traffic greater than any rail system. By the end of the sixties, as mass transit was coming into vogue, most American cities had already assumed the shapes and dispersions the highway generated. Three-quarters of the country's mass transit riders were in one city, New York. The rest of the country would have to depend chiefly on highways.

In response to criticism of their work, highway builders simply added more features. They would beautifully landscape the highways, they would hire members of minority groups, they would pay for relocation, they would build sound walls. They would provide parks and housing and replace the land they took. By the mid-sixties freeways had been reconceived as "urban development programs," bringing with them parks, schools, and housing. Two New York projects of the late sixties, the Cross Brooklyn Expressway and the Lower Manhattan Expressway, were touted as "linear cities," "megastructures" in which the highway itself was seen as an almost incidental element in the mix.

These "megastructures"—the term was fashionable among architects in the late sixties—were giant linear cities, like Le Corbusier's famous plan for a curvilinear Algiers. Paul Rudolph conceived one of these mad schemes for lower Manhattan: a highway, ensconced along with a rapid rail system, inside a tremendous roof composed of terraced houses. A "spine"—they always spoke of the megastructure's spine—of shopping and other facilities would run through the thing, from the West Side to Chinatown and the Manhattan Bridge, and perhaps over, into Brooklyn, like some self-replicating science fiction creature. It was to be, they said, "a new form of urbanism."

In truth, these projects simply represented all the bonuses that had to be added to expressways to make them at all palatable to the public, and in both these cases, the appeals failed: New York Mayor John Lindsay canceled the projects in the face of overwhelming opposition in the middle of his 1969 reelection campaign.

A few years later, another, perhaps ultimate megastructure project came up: Westway, a billion-dollar-a-mile project to replace part of the old West Side Elevated Expressway in Manhattan. The ingenious premise of Westway lay in its presentation not as a highway but as "Westway State Park"—a park that happened to have a highway beside or beneath it. To the usual inducements of jobs and contracts were added the public appeal of a park and the private appeal to real estate interests of the development opportunities near that park. And since the project was part of Interstate 478, the plan was a clever way to get the federal government to foot most of the bill for a new city park, a whole new city redevelopment program in disguise. To the Feds it was a highway; to locals it was a park.

Critics said the thing should never be built. The money should be "traded in" for mass transit, as allowed in the 1973 revision of the Highway Trust Fund. The trade-in was no sure thing, the proponents replied—even though New York was one of the handful of cities in the U.S. where mass transit made any sense at all—and it would disappear without a trace anyway into the maw of the New York subway deficit. It wouldn't leave anything *visible*.

If the planning of Westway proved that highways had come to be seen fundamentally as real estate projects, the demise of the project in 1985 reasserted the old principle that the federal government should initiate no internal improvement benefiting only a single state. The environmental issue that stalled and finally stopped the highway—whether landfill for the highway would harm the striped bass population in the Hudson River—was a mere pretext for opponents desiring to shift the money to mass transit. Considering the pollution of the river, it was almost laughable. But within the complex bureaucratic process highway building had become, the striped bass question was the fulcrum on which opponents overturned the whole massive project. The courts stalled it and Congress finally killed it.

The real reason for Westway's defeat was political: even Congress could not swallow the funding of so obvious a piece of porkbarrel, a project whose benefits were so completely local. This fact was pointed up by the jealous opposition of the New Jersey Congressional delegation, eager to pull some of Westway's potential real estate development across the river. Westway made a mockery of the principles of the Interstate program.

Moreover, both Congress and the public had come to sense a certain futility to building urban highways.

Everyone knew in advance that Westway—like all urban freeways—would soon be as congested as the rest of New York's freeways. It had long been clear that there was a sinister sort of logic at work in the world of traffic, a version of Parkinson's Law: the number of vehicles expands as if by magic to fill any highway, however capacious. Transportation expert Anthony Downs relates this law to the fact that, with approximately half as many cars as people in this country, no highway system could ever provide adequate capacity. Increasing capacity simply brings out more cars.

Not only did the roads fill up, they wore out faster than planned. It was as if the roads *created the traffic*—and in a sense, by spawning all sorts of development, by creating expectations through their presence, and even through their mere planning and projection, they did.

For the driver, the freeway offered an experience very different from driving on a wide-open, cross-country superhighway. The route tended to be one taken again and again, in common with many fellow drivers, and the choice making, the marginal changes in signs and perspectives, was not unlike the channel changing of television. The freeways forced individual choices of driving strategy, lane change, and exits, a logic of choice grown solid in the Boolean algebra of intersections.

Since so many of them are elevated, the driver on most urban expressways experiences a sensation of floating, of aerial rather than terrestrial movement, and a wide perspective that thrills the ago-raphiliac heart. The freeway seems constructed for the purpose of offering a safe view of the city. It may also foster racial and ethnic segregation. The driver can look down into the slums like a visitor to one of those "natural" zoos that run monorails among the lions and tigers. The distant skyscrapers become spectacular sculpture expressly for his delectation. Los Angeles and Houston seem to have felt compelled to create their skyscrapers not only to make themselves into "real cities," but to provide the driver constant visual evidence of that fact.

The space underneath the elevated highways, of course, quickly becomes a wasteland. Optimistic federal planners once touted its cold, dark shadows as fine places for parks and community markets, but few have ever appeared. More often, cars are abandoned there and garbage piles up. The city abhors a vacuum.

To the odologist, the urban freeways play the greatest role of any type of highway as public space. There is a certain commu-nication among drivers, a whole etiquette of merging and diverging, a routine of eye contact and self-presentation in which the aesthetics of the individual vehicle itself naturally plays a role. Los Angeles and Houston abound with tales of love matches inspired by the look of a fellow driver's car—often the immediate impetus to the meeting is a fender bender. The freeway driver, one half of his brain occupied by the immediate and mostly automatic decision making of driving, is, like the television viewer, in a peculiarly distracted and suggestible state of mind. Unnatural as they are, not even sharing the shape of the terrain like rural superhighways, the wide open spaces of the freeway are nonetheless a sort of modern equivalent of Romantic nature. Thus Reyner Banham argues that coming off the freeways in Los Angeles is like coming inside.

The freeway is a place where one can lose oneself. Connoisseurs of Los Angeles anomie and alienation like Joan Didion and her husband John Gregory Dunne have made this set of freeway sen-sations legend. "I went out for a loaf of bread and ended up in San Francisco," Dunne once claimed. Didion's description of her her-oine taking satisfaction driving on the freeways is the classic exposition of the "existential" experience of the urban expressway: "Again and again she returned to an intricate stretch just south of the interchange where successful passage from the Hollywood onto the Harbor required a diagonal move across four lanes of traffic. On the afternoon she finally did it without once breaking or once losing the beat on the radio she was exhilarated, and that night slept dreamlessly."

For Didion's heroine, Maria Weymouth, driving the freeways is therapy. Like many drivers she finds on the freeway a sense of something controllable, a realm where a portion of the mind is engaged in partly mechanical, partly intuitive actions, where volition is reduced to minimal levels.

Part of the freeway is about losing yourself in the crowd. People who could not stand to be on a bus or train full of strangers find merging their vehicle with the flow somehow refreshing, almost baptismal: a form of going down to the river.

But the freeway also inspires paranoia and phobia, an overwhelming sense of powerlessness that compels some drivers to pull off on the shoulder or simply stop in midlane. A Texas psychologist began testing the victims of these disorders, but found a problem: most were too frightened to drive across town to his laboratory.

Radio reports of traffic conditions, relayed from helicopters and planes, and the work of computerized traffic control centers aim not at control of traffic—except for the damage control of removing disabled cars—but simply at issuing warnings. Their work is like that of the weather service: everyone talks about traffic but no one ever does anything about it.

The same place names naturally recur in the reports from day to day—the Stack and Spaghetti Bowl in Los Angeles, the Mousetrap in Denver, the intersections, bridge and tunnel entrances, and other bottlenecks—making them among the most notable public landmarks in the common imagination of the city, places around which swirl the storms of rush hour.

Seen from the helicopter or on the remote video screens of traffic central, breakdowns and accidents are facts of nature, like hurricanes picked up by spotter planes or in satellite photographs. From the perspective of the helicopter, individual driver action appears as a simple statistical fact. The anger or sympathy directed at a stalled motorist by his fellows on the ground is absent. The chopper pilot is generally impressed that the highway system works as well as it does. Unfortunate incidents seem inconsequential to the pattern of the whole, and the aerial perspective occasions thoughts not just of the smallness of the individual automotive corpuscle but of the beauty of the collective flow. As always, the rule of odology holds which states that viewing the pattern of the highway from above mystifies and romanticizes it.

The most famous freeways, of course, are those of Los Angeles. A freeway view, often with palms and the Hollywood sign placed in the background by artistic license, became as much the city's trademark image as the Empire State Building for New York or the Eiffel Tower for Paris.

There is a myth that the freeways "created" the classic mobility associated with California. This is the old error of technological determinism. It is true that there are more cars than people in Los Angeles, but that is a result, not a cause. Los Angeles was built as a far-flung city by people who treasured mobility and distance between themselves and others. Its rail network, first that of the Southern Pacific and later that of the local Pacific Electric, ran to well over a thousand miles and covered virtually the same territory as the freeways system, whose routes in many cases paralleled the rail routes.

Freeways are a much later arrival on the Los Angeles scene than we generally think. Their building began with the Arroyo Seco, still called a parkway on its completion in 1940 and later renamed the Pasadena Freeway. It got going in earnest only in the late fifties, particularly after Interstate money became available.

Other cities had pioneered the freeway concept. New York, with the Westchester and Long Island parkways substantially completed before World War II, was the leading model. Detroit had laid the first hard pavement outside the city limits, in one of its periodic extensions of Woodward Avenue, in 1908, and planned a super-highway system, although its routes were simply to be wide thoroughfares, without limited access, in the twenties. Chicago's Lake Shore Drive had combined traditional parkway design with the more efficient intersections of superhighways by 1905.

New York's parkways were perhaps the most influential model; even today, New York—of all cities, the one least associated in the public mind with freeways—has more miles of limited access highway per square mile than Los Angeles or Houston.

Los Angeles planners, led by city engineer Lloyd Aldrich, visited New York's parkways in search of models for their own traffic needs. The vision of a Los Angeles "parkway system" dates back at least to 1939, when a series of 613 miles of parkways was proposed by Aldrich after a study financed by such business leaders as P. G. Winnett, the head of Bullock's department store. The Arroyo Seco was the only prewar product of this plan. War and growth changed the parkways to freeways: by 1947 a good portion of the Hollywood Freeway was complete.

It is true that Los Angeles, and California in general, also saw the birth of two elements fostering road building that were to be reproduced in larger scale at the national level. California was one of the first states to adopt a law allowing the state to limit access. (Access, a property right, had to be acquired by eminent domain.) It also pioneered the "trust fund" concept, created by a constitutional amendment in 1938, with the specific reservation of road tax funds for highway construction, that was to become a model for the financial engine of federal roadbuilding after 1956.

The state also possessed a legendarily strong road lobby, led by State Senator Randolph Collier; its combination of oil, automobile, rubber, and construction companies was to be repeated in the national road lobby.

Collier, from the rural district of Yreka, in the Siskiyou Mountains, was a curious figure. The "father of the freeways," "the silver fox of the Siskiyous," creator of the Collier-Burns Act, was always described in the press as "white-maned." He dominated the state's road program for nearly forty years and used it as a source of tremendous political power.

"Randy's Ratpack," as the highway lobbyists were known, had free run of the state house and almost total power to designate road projects while Collier was head of the transportation committee. He rammed freeway bills through and took pride in tagging himself "the fastest gavel in the West." He defeated repeated efforts by San Francisco and Los Angeles to have some highway funds diverted to rail systems by ridiculing it as "rabbit transit."

But the main reason Los Angeles was associated with freeways was that freeways implied values associated with that city. Los Angeles summed up the myth of a carefree, sunny, showy life that Americans associated with the freedom of the road; thus the legends of families living in mobile homes perpetually revolving through the freeways (the Okies permanently caught up in the process that brought them west), of auto-mated prostitutes, of literally floating crap games in vans on the freeways.

It was the end of all migrations, a "lifestyle" of competition, big money and fast money, speed and glamor—the way of life that gave birth to the metaphor of "the fast lane."

Indeed, it was not until the arrival of the freeways that the city began to take on an identity as a city in the national mind. Before that, people had thought of the movie stars living in Beverly Hills, of the studios in Culver City, or the myths of Hollywood. They associated the city with discrete places that were part of it; everyone knew that Los Angeles "had no center." It was the freeways that linked these separate communities. In 1960, a *Life* magazine survey of the city referred to the "ribbons of freeway which are gradually tying the city's scattered pieces together."

And yet the freeways also held the pieces apart. By contrast with the New York system on which they were modelled, they worked because the city was so spread out. In the 1970s, when the city, long irked by the feeling it had no downtown, began to acquire the cluster of skyscrapers that passes for downtown in most American cities, its traffic jams for the first time became chronic.

As the dominant conception of superhighways in the city was one of escape, it was fitting that in the city where they were most important they were named for destinations—Pasadena, San Diego, Santa Monica. All freeways led *from* Los Angeles.

· · ·

The freeways supported, but they did not create, the shape of Los Angeles. They did create the shapes of Houston, where post-cards proudly assemble views of major intersections, and of Atlanta and a host of smaller cities. The pattern that developed in these cities, which enjoyed their major growth only after the advent of the Interstate system, was that of the "orbital" city. The peripheral bypasses the Interstates engineers drew, based on their traffic sur-veys, became the freeway Main Streets of Houston, Atlanta, and dozens of smaller cities. Even Washington underwent its critical growth spurt in this period and on this pattern. The limits of the District of Columbia gave way to the Capital Beltway as defining the range of the federal influence. Seasoned politicos began to speak of their constituents as those "beyond the Beltway."

On the great orbital freeways, the centripetal attractions of jobs and entertainment in the city's center were balanced out by the centrifugal attractions of suburbia and ruburbia beyond. You could get downtown or out of town by radial connectors, and you could get around town by the beltways. As Jefferson and Gallatin had planned roads to preserve the country from the pestilence of cities, the orbitals helped save the city from itself, its own congestion and centralization. Locations along the peripheral freeways supplanted those in the central city not only for residential developments and shopping centers but for office and light manufacturing complexes. They were parodies of the "greenbelt" visions of optimistic city planners, circumferential main streets that, rather than marking the city's edges, became its skeleton. As congestion began to choke even these belts, wider ones were planned, like Route 495 which supplants the I-95/Route 128 Beltway in Boston, and the possibility was raised of cities growing concentrically outward, the succession of their orbital highways marking the stages of their growth as clearly as the rings of a tree.

Popularizing "Magic Motorways." Norman Bel Geddes's vision of a super cloverleaf, from the General Motors Futurama model at the 1939 World's Fair. (GENERAL MOTORS)

TWO
Shaping the Flow

Mass producing highways.
Construction of the
Pennsylvania Turnpike.
(THOMAS McAVOY, © 1949
TIME INC.)

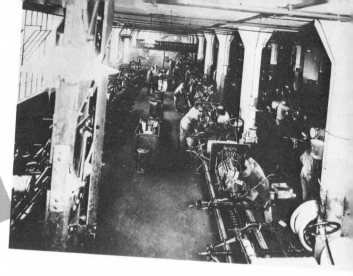

Mass producing automobiles. Reinforced concrete columns and assembly line at Ford's Highland Park factory, designed by Albert Kahn. (FORD MOTOR COMPANY)

7: *Streamline to Assembly Line*

You sat, partially enclosed, in a blue velour seat intended to give the illusion of airplane flight over "the America of 1960." There were 322 of you, forty-five tons of human cargo, swept along for a fifteen-minute "magic Alladin-like flight through time and space," as the whisper of a recorded narrator called it. You gazed down at models of 500,000 buildings, among which streamed some 50,000 teardrop-shaped cars, and a million trees—made of copper wire and Norwegian moss. A machine called the Polyrhetor carefully synchronized that low, intimate narrator's voice—the product of multiple loops of movie sound film—with the movement of the seats.

As you crossed America, the rosy dusk of the dioramas turned into night, the better to show off the lighting of the landscape, and then, as the "plane" moved across the country, past a new Metropolis based loosely on St. Louis, the sun rose again.

But the highlight of the models was the superhighway system: five different lanes in each direction, filled with the tiny, teardrop-shaped cars coursing endlessly, some at 50, some at 75, some at 100 mph, crossing from one great freeway onto another by wide, swooping ramps. The whispering voice talked of crossing America by car within a single day.

Lights embedded in the rails of the highways automatically switched on as automobiles neared them; bridges featured traffic on different levels at different speeds. City and country, industrial and residential areas were neatly divided. Each city block of the Metropolis of a million souls, stacked with Moderne skyscrapers, is "self-contained," while most workers lived in planned suburbs.

At the end of the flight, you stepped onto a rotating circular platform that shunted you to a life-size replica of the intersections in the models: pedestrian bridges over a wide, multilane street for automobiles.

It was, of course, the Highways and Horizons exhibit at General Motors' Futurama: the world of 1960 as seen at the World's Fair of 1939. The Futurama exhibit had more impact than any other single event in acquainting the wider public with the possibilities of superhighways. It was the most popular attraction of the New York World's Fair; as many as 10,000 people might wait two hours or more on any given day to take the fifteen-minute ride. Ten million people eventually visited the 44,000-square-foot, six-million-dollar diorama. The Futurama, as one astute newspaper reviewer put it, combined the "thrills of Coney Island with the glories of Le Corbusier."

The exhibit was designed by Norman Bel Geddes, one of the most flamboyant designers in an era that saw the industrial designer emerge as a public figure, a celebrity, a virtual folk hero of industry. Raymond Loewy, Henry Dreyfuss, and Walter Dorwin Teague were, along with Bel Geddes, the best-known of these.

All were represented at the Fair: Dreyfuss with his Democracity, inside the Perisphere; Lowey, with the Transportation Zone at the Chrysler Building; Teague, with the giant cash register that stood atop the National Cash Register building, and the Ford pavilion, where visitors could test-drive an automobile along a half-mile, spiraling "Road of Tomorrow."

The common aesthetic of these new designers was streamlining—the shaping of nearly every sort of object as if it were required to push through the air like a plane. Streamlining had become a popular value. Its overtones extended far beyond the fluid aerodynamics which had created it. Its purely aesthetic appeal was illustrated by pencil sharpeners and furniture designed with "streamlined" forms largely inspired by the airplane, the technological darling of the age. But streamlining as a popular concept implied more. It meant leaving the roughhewn American past behind, it meant metal and plastic instead of stone and wood, it meant urbanity instead of homeyness. It was the form of the bright future that must lie somewhere beyond the sufferings of the Depression.

Streamlining manifested itself in highway design in the new science of traffic flow: the science of keeping the open road open. The parkway had already shown how to reduce the friction of traffic entering the road by limiting access. Superhighways would streamline flow by separating traffic at different speeds and shaping curves for maximum speed.

Poll-style surveys and mechanical counters were developing traffic science into an organized discipline and creating a sound and quantitative sociology of automobile movement. To manipulate the results of their surveys, traffic engineers began to borrow equations from hydrodynamics.

A man named Miller McClintock, Director of the Erskine Bureau of Street Traffic Research at Harvard, funded by Studebaker, had developed the "friction" theory from this sort of data. McClintock argued that accidents and traffic jams were caused by several sorts of friction: from traffic entering or leaving the road, from slow-moving vehicles, or from intersections. The answer to safer, faster, more beautiful roads was to eliminate such friction, and Bel Geddes had shown how.

Streamlining became an economic metaphor as well. The Depression, it was thought, was the result of a "sticky" economy, burdened by "friction" that kept consumer demand and production from moving smoothly together. Henry Ford had anticipated such thinking when he raised the pay of his workers to $5 a day—not out of simple generosity or with an eye to better labor relations but so they could afford to purchase the cars they were building. Ford called it "efficiency engineering." Now, proponents of "consumer engineering" argued for the need to properly package and advertise products in order to end the "underconsumption" they believed kept production and demand from moving smoothly together. It was a pseudoscientific version of the famous GM executive's lament, "It's not that we make such lousy cars, it's that they are such lousy consumers."

Consumer engineering meant designing products with an eye to the buyer, not to the efficiencies of manufacture. Consumer engineering was the annual model change of Alfred Sloan, as opposed to the one-product engineering of Henry Ford's Model T. While the annual change built obsolescence into cars, the marketing of a series of models—Cadillac, Pontiac, Buick, Oldsmobile, and Chevrolet—created a social hierarchy of styles, what Daniel Boorstin termed a "ladder of consumption," that consumers could aspire to climb. And as the car was redesigned for the consumer, consumer engineering also came to mean designing and building highways for *their* consumer, the driver.

In 1937, McClintock had extended the streamlining he proposed for highway design to economics in a way that must have pleased the brass at Studebaker and General Motors. If streets and highways were made efficient—elevated freeways injected into the city, superhighways built through the country—he said, then Americans could buy five or ten billion more automobiles.

> The consumption and use of automobiles at the present time is controlled not by the ability of the American people to buy more automobiles but by their inability to use such automobiles as they have, or might have, with a decent degree of security and convenience.

Consumption, like traffic flow, was limited by the friction of bad roads. The best way to "engineer" higher consumption, therefore, was to engineer new highways.

Consumer engineering fitted nicely into a Zeitgeist where jazz and the DC-3 airliner, stream of consciousness in literature and the Chrysler Airflow automobile, all seemed to blend in a grand vision. Bel Geddes's superhighways were part of the vision: they would make the economy, indeed the whole society, flow more openly and smoothly.

Bel Geddes's greatest strength as a designer was as a propagandist. Perhaps this was a legacy of his beginnings in theater design and advertising. To stage *The Miracle* in 1923, he turned the Century Theater into a reproduction of the interior of a great cathedral. He designed operas for the Met and productions for the Ringling Brothers' Circus.

But most famous were his visionary projects: the Aquarium Restaurant, to be built inside a dam, where diners looked through an illuminated waterfall, or the Revolving Aerial Restaurant, later imitated in the Seattle Space Needle and various of John Portman's Hyatt Hotels.

But he was not above designing such frivolities as the "Crystal Lassies" show at the Fair, where scantily-clad women were reflected for an audience in a complicated set of mirrors, or dreaming up a fantastic proposal for a new Brooklyn Dodgers stadium that would feature remote concession service through the arms of each seat.

He proposed streamlined ocean liners and trains, patented several "teardrop-shaped" automobiles, and offered the vision of a giant 400-passenger airliner shaped like a huge flying wing. He created the modern style of airliner interior, with fold-down trays for food service and zippered fabric covers, for Pan American's early Clippers.

Born Norman Geddes, he had even redesigned his name, adding the flashier, classier "Bel"—borrowed from his first wife's middle name—to smooth out its rhythm and proportions. (His daughter, the actress Barbara Bel Geddes, retained the odd spelling.)

He had moved into design through advertising. One of Bel Geddes's advertising clients was Stanley Resor of the J. Walter Thompson Agency. The two got along well and eventually Resor's niece, Frances Resor Waites, became Bel Geddes's second wife.

Resor was a man of progressive tastes, who was responsible for Mies van der Rohe's move to the United States. At Philip Johnson's suggestion, he hired Mies to design a house for him in Jackson Hole, Wyoming. The house, whose architecture was even more minimal than usual in Mies, emphasized the dramatic view of the mountains. Although the house was never built, Mies stayed in the United States. Resor helped link Bel Geddes up with Shell Oil, for whom in 1937 he designed a model city, packed with freeways, called Metropolis. It was a virtual study for the Highways and Horizons exhibit. Bel Geddes tried to interest Shell in expanding Metropolis for the Fair. When it declined, he went to GM.

Underneath Bel Geddes's showmanship, however, lay a blend of traditional American functionalism—Horatio Greenough's clipper ship—with the emerging dictates of European modernism. Steamlining was positive because it marked the happy and necessary coincidence of function and beauty in a formula that could be proven mathematically: streamlining produced efficiency and higher speeds in moving vehicles. Extension of this principle was by metaphor only, but that did not stop it from reaching into architecture, the design of immobile products, and even clothing design.

Bel Geddes praised the requisite list of objects of good design favored by "progressive" designers of the time: the clipper ship, the ocean liner, the airplane, the dynamo (represented in his book by photographs by Margaret Bourke-White), the grain elevator, the concrete dam. He compared the beauty of highways to Brancusi's sculpture.

He viewed the airplane as a paradigm of good design because it was the clean, simple approach to a fresh problem. Contrasting the airplane with the automobile and its "horseless carriage" ancestry, he wrote that aircraft would have looked far different if designers had approached them as "winged carriages."

But what was critical was his addition of "social design" to the list of design principles. Writing in 1932, he had gone so far as to praise "the great experiments of Russia" in centralization as examples of social design. And Bel Geddes's Futurama narration invoked the ideas of consumer and economic streamlining: "A freeflowing movement of people and goods across our nation is a requirement of modern living and prosperity."

"We are entering an era," he wrote in *Horizons*, "which, notably, shall be characterized by *design* in four specific phases: design in social structure to insure the organization of people, work, wealth, and leisure. Design in machines that shall improve working conditions by eliminating drudgery. Design in all objects of daily use that shall make them economical, durable, convenient, congenial to everyone. Design in the arts, painting, sculpture, music, literature, and architecture, that shall inspire the new era." Highways offered a combination between design of an everyday object and social design: Bel Geddes's "magic motorways" would reshape the landscape, and thereby reshape society.

"Social design," for Bel Geddes, among other things meant reshaping the cities so that people could work downtown and yet live in the country. Close to the heart of Bel Geddes's highway schemes was a vision of suburban life, but suburban life more grandly justified than ever before.

As we have seen, his magic motorways would extend the commuting radius six times. It would no longer be a matter of the distance between home and office or factory, but the time between them. After all, this was the age of Einstein, when space and time had been discovered (at least this was what the public thought it understood about relativity) to be one and the same, inseparable and interchangeable.

There were a few problems with the Futurama exhibit. Occasionally, one of the tiny cars on the motorways would get stuck in its groove and tens of other cars would crash into it, creating a huge, miniaturized, chain-reaction crash. It was a distant hint that all might not run quite as smoothly on the highways of 1960 as Bel Geddes claimed.

But the device that moved the people through the exhibit performed flawlessly. The shape of the pavilion itself—designed by Albert Kahn, creator of the great auto plants for Packard and Ford— kept the crowd flowing along a series of streamlined ramps to and from the chair train. And to carry viewers over this ultimate vision of individual automobile travel, Bel Geddes and Westinghouse engineer James Dunlop had, ironically, created a miniature mass transit system. It was like an assembly line, a factory for the education of the public in the glories of superhighways.

Cinematic architecture. The Linn Cove Viaduct,
1984, part of the Blue Ridge Parkway. The
highway floats over the landscape on prefabricated
supports, creating sweeping views over the
surrounding valleys of the Great Smokies.
(PRESIDENTIAL DESIGN AWARDS PROGRAM)

8: *Mies en scène:*
The Shape of the
Superhighway

"I love the concrete, my brother!"

—*Horatio Greenough*

"As new and greater road-systems are added year by year they are more splendidly built. I foresee that roads will soon be architecture too . . . great architecture."

—*Frank Lloyd Wright,* An Autobiography

There is something curious about photographs showing Model T's and Model A's being assembled at Albert Kahn's Highland Park Ford factory. Above the cars, in what seems to the odologist a striking contrast in technologies, stand futuristic-looking concrete mushroom columns. Beside them, the cars look almost medieval; the sweeping beams and columns could be part of the latest Interstate interchange.

The similarity suggests something fundamental about the superhighways: that they were like factories for transportation, assembly lines generating the "trips" that highway engineers talked about as their basic "product."

Kahn used concrete construction to keep the floor space of Ford's factories as clear as possible and the assembly lines flowing. The factories were vast. Everything from steel to glass was manufactured from raw materials on the huge site. The different streams of materials and subassemblies all flowed together into the final product—a complete industrial integration idealized in Charles Sheeler's precisionist painting of River Rouge, with its ore barges and railroad tracks.

As Kahn's factories lay somewhere between architecture and engineering, the best of the superhighways occupy an area between engineering and landscape design. They are designed according to the same claims, partly competitive, partly congruent, of cost and aesthetics that David Billington lists as the bases for "structural art."

Kahn designed his factories, beginning in 1903 with a Packard plant when there were no more than a thousand cars on the streets of Detroit, in this form not because he thought them especially beautiful but because they were functional. Kahn's architecture for non-factory buildings and even for the office sections in front of the factories was quite conventional, an application of classical elements.

The concrete post and slab construction Kahn used in his auto factories had been developed in Europe around the turn of the century and pioneered in the United States by C.A.P. Turner. But it was from Kahn's models, ironically, that many of the modern European architects came to appreciate it.

Accompanying the concrete was a sense of openness in the wall. The Highland Park plant had 50,000 square feet of glass for illuminating the work area. The glass gave the structure, which was 865 feet long, a lightness that defied its bulk.

To the European modernists, the flow patterns of Kahn's factories achieved for functional reasons exactly the flow of space they sought for aesthetic ones. Kahn, a German emigré, was a darling of the European modernists. Gropius's design for the Bauhaus, with its vast window walls of glass and its concrete columns, was an almost literal translation of a Kahn factory. Mies van der Rohe, who worked for Peter Behrens, the great factory designer, and who built roads for the German army during World War I, piled Kahn-like structures one on top of the other for skyscraper and department store designs for Berlin. And Le Corbusier planned freeways for his ideal cities on columns like those that supported his "machines for living."

Driving on the modern superhighway you saw the factory, as it were, from the assembly line itself: from the point of view of the product. The result was a structure that revealed itself in motion, cinematically, and yet, in its complex interchanges, remained a mystery. It was no easier for the driver to quite put together in his head the structure of an interchange—the mixture of its columns and ramps, its openness and enclosure, its demand that you turn right in order to turn left—than it would be for the moviegoer to tell you as he left the theater how the tracks had been laid for the camera movements.

The modernism of the superhighway was the modernism of the transported eye, of flowing sight as well as flowing space. The parkway had appeal because lovely terrain filled the screen the highway projected, providing a scenic backdrop. The superhighway shifted backdrops rapidly, turning distant landmarks into actors on the wide windscreen.

To the people Kevin Lynch interviewed for his *The Image of the City* (1960), the highway was "a succession of events" whose arrangement, Lynch hypothesized, possessed a "melodic order" that could be composed by a skillful designer. But it was to the individual driver or passenger that this order unfolded itself. The individual viewer in his automobile was the center of the action: the car was the star.

The highway designer could actually realize what the modernist architect most often could do only metaphorically: model form on function. For the highway, of course, the function was that of movement. The design speed of the highway was the basis of its entire architectural "program," and the intersection was the natural locus of interpenetrating planes. (As speeds went up, the changing program produced a road structure that was far more different from the original two-lane motor highway than that highway was from the wagon road.)

In this, road designers echoed the principles of modern architecture. For all the differences among the masters of modern architecture, they agreed on one thing: the open flow of spaces. Wright might inveigh against the boxes and "drawing board architecture" of the Europeans, but a Wright house shared with Mies's Barcelona pavilion the sense of interpenetrating spaces.

But the best highway designers also looked back to a native American tradition of functionalism that enjoyed a revival of attention in the 1930s and 1940s, just as the superhighway was being conceived. It was a revival that took place partly in the light of European modernism—Le Corbusier, for instance, had praised the power of functional expression in American bridges, warehouses, and grain elevators.

The revival included the rediscovery of Horatio Greenough. Greenough was better known during his lifetime for his neoclassical sculpture, including a controversial statue of a nearly nude George Washington, than for the writings on functionalism which were championed by Lewis Mumford, John Kouwenhoven, and others and are now recognized as one of the crucial documents of American aesthetics.

Greenough argued that any design that expressed function efficiently and simply was necessarily beautiful. He held that popular approval and use were the highest standard for art, both fine and practical. He praised the "majesty of the essential" and offered as models the design of the clipper ship, along with American fortifications, bridges, and other purely "scientific creations," in contrast to the "imitation of the Greeks" that was the characteristic quality of the architecture of his time—and indeed of his own sculpture.

"Scientifically" created to maximize the performance of the modern automobile, shaped for speed as carefully as Donald McKay's *Flying Cloud* clipper ship, streamlined free of roadside interference, the modern highway realized Greenough's ideals.

Another inspiration for the best of American superhighway design was the precepts laid out by two of the youthful designers of the Autobahns: Hans Lorenz and Fritz Heller. Lorenz, in particular, stood in relation to highway design much as Wernher von Braun stood in relation to rocket design. As a young engineer working on the Autobahns, like the 1940 Breslau-Vienna stretch, he came up with highly advanced theories, applied in full after the war to the new generation of Autobahns, notably the much praised Anschaffenburg-Nürnberg stretch.

Most previous highway design had seen the road as a series of straight stretches—known to engineers as tangents—tied together by curves—called arcs. Lorenz's contribution was to blend these separate elements into a series of spirals (technically, helixes) to produce a steady, continuous flow without abrupt jumps into curvature. His roads were designed for—and designed *to*—a steady, consistent speed.

Lorenz argued that curves generally cease to be visually significant to the driver if the visible part of the curve accomplishes a turn of two degrees or less—or, in engineer's talk, only 7.6 feet deflection from the straight line over a 50,000 foot radius.

His theories combined the horizontal elements of highway design—the tangents and curves—with the vertical elements—slopes and levels. The Lorenz highway was a three-dimensional model for which definite geometric prescriptions could be written. It was also a system that led to easier blending of the highway into the landscape, with fewer sharp-walled cuts, and harmonized with the Nazi ideal of making the highway as natural as possible a part of the geography. It was a typically German resort to mathematics to combine efficiency with aesthetics.

These principles, while applied in part by Gilmore Clarke and others to parkway design, entered mainstream American highway design only after World War II, and only slowly. The doctrines of American highway building before World War II were summed up in the dicta of Charles Upham, one of the country's most respected highway designers. Arguing that commerical roads, whose value lay in the straight line, were absolutely to be distinguished from scenic or recreational roads, where curves had aesthetic value, Upham laid out a geometry of maximum tangents, where the straight line was abandoned only when absolutely necessary, and for a curve of minimum length. The driver was expected to vary his speed for the curve, not the curve to vary to assist the driver. Thus the profusion of signs on American highways of even the most "advanced" type bearing figures of curves ahead and an injunction to cut speed from the maximum.

The idea of a steady, intended speed as the basis for design was a German idea that American designers only grudgingly accepted. When they did, they claimed to have invented it. A footnote in a highway history prepared in 1976 by the Department of Transportation still credits the Germans only with "concurrently" developing this idea.

Until the fifties, American thinking continued to value the straightest highway possible, with curves built into it only under the duress of landscape. It ignored mounting evidence that straight roads were boring to the driver, and because boring, more dangerous than those whose curves kept the driver involved and, at best, entertained. Curves were thought to be the source of accidents. Early literature promoting the Pennsylvania Turnpike, in fact, boasted of one stretch west of Carlisle that ran for nearly forty miles with only one slight bend.

By the time the Interstate program began in earnest, however, it was recognized that the straight line was the highway designer's greatest enemy, in terms of both safety and aesthetics. The greatest complaint about the superhighway is the tedium of driving on it. The straightest, most boring roads, like great stretches of the Kansas

Turnpike, have some of the highest accident rates in the country. Those rated as aesthetically attractive, blending curves and straight-aways, are also among the safest.

Dull highways can literally put drivers to sleep at the wheel. They also reduce their sense of distance. Designers speak of "velocitation," the tendency of drivers too long on a straightaway to underestimate their speed. The results are rear-end collisions, sportscars sliding underneath tractor-trailers, and chain-reaction crashes.

But the same factor that killed the functionalist Model T—the human desire for variety and expression exploited by GM's annual style changes—was the source of the common criticism of "modern functionalist" architecture and the "boring, impersonal Interstates." The driver on the Interstate could quickly feel as alienated as the worker on the Model-T assembly line. And almost as soon as the Interstates and other superhighways were completed, as soon as the novelty of these roads of the future began to wear off, travelers were complaining that they offered no views of the real countryside, that they were dull.

The superhighway is distinguished from the traditional roads it meets at intersections by its bridges. The old roads pass over or under the superhighway, never touching it directly. Planes shift and interpenetrate at superhighway intersections like the elements of a modernist house. In the bridge, the superhighway has its own particular architecture.

The superhighway bridge is not the soaring bridge of poetic reveries, arching higher than any construction in the city, the Brooklyn Bridge or the Golden Gate, or even the virtuoso reinforced concrete of Robert Maillart's Swiss bridges, individualistic works, rightly treasured by the modernists, but each as unique as a Le Corbusier building.

While the aspects of bridge construction that inspired Whitman and Hart Crane were those of verticality and often mystic con-struction, the bridges of the highway stressed horizontality and extension. You went under as well as over these bridges. Neither level was privileged. They proposed to join themselves as nearly as possible with the landscape and the road, acting as an aesthetic as well as physical junction between the two.

The bridges of the American highway had to be mass designed and adapted to the variants of their sites. Technically called "grade separation structures" by the engineers, they were like little sections of Kahn factories. By the beginning of the eighties there were an estimated 500,000 highway bridges in the United States.

The overpass bridge had its origins in canal and railroad building. In 1830 Colonel Stephen H. Long, an Army engineer, constructed a truss bridge to carry the Baltimore-Washington wagon road over the railroad, perhaps the first overpass in the country. But the real ancestors of the overpass were Olmsted's park bridges. Olmsted included the first reinforced concrete bridges in his designs for Prospect Park in Brooklyn in the 1860s. For years, such bridges were faced in rusticated stone.

For the Westchester parkways, engineer Arthur Hayden produced a virtuoso series of rigid boxframe overpasses that were widely imitated. They were showpieces of good concrete design, cast in a single piece and relatively inexpensive. Most of the parkway bridges, however, were sheathed in masonry for aesthetic reasons.

By the thirties, in metal bridges like those carrying New York's West Side Expressway, Art Deco had surfaced, in the form of the railings and in decorative medallions indicating the names of the ship piers that lay behind the highway. In the overpasses of the Pennsylvania Turnpike, the Brooklyn-Queens Expressway, or the original Arroyo Seco Parkway in Los Angeles, the boxy concrete bridges are softened with decorations in a sort of Post Office Moderne: rounded solids or bas-relief forms, noncommittally modern and capable of reproduction with little variation.

The design of bridges on the Interstates has seen an evolution toward lightening and opening the space underneath. The thrust of the change was toward clearing away the "tunnel effect" of older overpasses and establishing the equality of the two roadways. There was a brief fad in the sixties for tilted and dramatic support, but most of the change comes from extending the length of the bridge, carrying it over receding slopes of grass or concrete on longer beams. Often, steel girders are left exposed and, painted blue or green or brown, provide dramatic horizontal elements.

Everything possible was done to emphasize horizontality and flow. It became the industry ideal to lengthen and strengthen overpass bridges, even at additional cost, and to replace boxy or arched bridges, grounded in the earth, with longer ones where the sides were cut back on an angle. Above all, the tendency was to reduce the number of columns in order to make the structure seem to float.

While some of the benefits of these designs were practical—fewer columns meant fewer collisions with them, wider spans meant space for future expansion and easier maintenance—the main thrust was aesthetic. The new bridges corresponded to a vision of what driving should be like. "Fewer and longer spans give drivers a sense of openness which contributes to relaxed driving," said a 1963 Bureau of Public Roads memorandum, summing up the new way of thinking for local engineers. "These newer designs reflect the dynamic feeling of traffic movement, with emphasis on long horizontal lines uninterrupted by vertical elements."

This sort of policy edict recognized that in spite of those famous interchanges where bridges were piled three or four atop each other, the dominant lines of the highway were naturally horizontal. There was in America an almost mystic reverence for this horizontal. It was, after all, the line of the prairie, of the beckoning horizon, of the open land untamed by church steeple or courthouse cupola, to say nothing of the skyscraper. It was the line of destiny manifested.

For Frank Lloyd Wright, this horizontal line, around which he organized his prairie houses, was "the line of freedom." "I see this extended horizontal line as the true earth-line of human life, indicative of freedom," he wrote.

> The great highway is becoming, and rapidly, the horizontal line
> of a new Freedom extending from ocean to ocean, tying woods,
> streams, mountains and plains and great and small together, tapping

an outflow from the overcrowded urban fields for the better building
of men in a better way.

The horizontal line of the road was echoed in the increasingly
horizontal shape of the automobile itself, the "long, low, lean look"
that General Motors in particular developed. Harley Earl, the first
of the great auto designers, creator of the Cadillac LaSalle, the
Pontiac Firebird, and other cars, endorsed the horizontal look. It
was, he said, the look of the ranch house or the great factory, and
the American people liked it.

The beginning of automobile styling in the late twenties marked
the gradual separation of the car from the road. Automobiles, from
the streamlined Chrysler Airflow to the tail-finned Cadillacs and
other models introduced from the late forties on, expressed an
aspiration to the role of the airplane. The body styles rendered this
sort of design symbolically; their exaggerated suspension, power
brakes, power steering, and automatic transmissions—with trademark names like "Aeroglide"—expressed it functionally. The designers, and the drivers who bought their products, wanted cars
that floated on air, unattached to humble pavement. Such vehicles
seemed appropriate to superhighways that themselves floated
through the landscape, unslowed by the rise and fall of the hilly
countryside, on platforms well above the sooty miscellany of the
city.

If the bridges and viaducts suggest a genre of architecture, then
the earthworks of highway cuts and fills, the reshaping of land and
the insinuation of the highway into its forms, suggest an art of their
own. Artists like Michael Heizer, Robert Smithson, Carl Andre,
and others went to the Western deserts to slice and frame the earth,
to absorb the energy of the wide, open spaces into their structures.
The highway, cutting across desert or plain, through mountainside
or valley, created a similar effect.

The parkways had first shown the power of this sort of design.
As Olmsted moved thousands of cubic yards of earth to shape the
"natural" contours of his parks, providing a theatrical version of

romantic landscape, so the parkways succeeded in shaping the drama of terrain and vistas by location and careful screening or framing. In the parkways designed by Gilmore Clarke, Clarence Combs, or Stanley Abbott, each roadway was manipulated independently, with the median widening into a hilly island or long, grassed meadow.

It was remarkable just how little land was required around a parkway to give the impression that you were driving in the middle of a vast forest or amid lush farming territory (when, in fact, most of the highway was surrounded by suburban housing tracts). William Whyte has noted how one or two dramatic vistas can color the motorist's impression of the whole route. On the Taconic, he points out, "there is one particularly striking view as you emerge from a forested section and see before you a sweep of rolling meadowland with old farms and barns and silos. It is these intermittent Grandma Moses scenes that people remember; and though in total there are only a few miles of them, in the mind's eye they set the character of thirty or forty miles."

On the Natchez Trace Parkway, where the right of way is sometimes as narrow as a hundred feet, the motorist proceeds through what seems like a piney, Scandinavian park and passes over a narrow country road, offering a sudden glimpse of backwoods, Faulknerian Mississippi: red clay shoulders and rusting, tinroofed shacks. Along Route 1 in California, south of Monterey, there is a particular curve, much photographed for travel periodicals, that frames a rocky headland with crashing waves and embraces the quiet concave curve of a beach. Such realizations show the potential subtlety of the road designer's art, his manipulation of *mies en scène*; and the best of the superhighways incorporated it.

The highway also set up patterns to be read from the air and from a distance on the ground. Long after the pavement is gone, the four-hundred-foot cliff of the cut on I-40 along the Pigeon River in North Carolina or the huge slope of the Interstates in the Rockies will survive as features of the land.

But the superhighways, as their admirers have noted, were themselves sculpture. Reyner Banham argued that the intersection of the Santa Monica and San Diego freeways in Los Angeles is "a work of art, both as a pattern on the map, as a monument against the sky, and as a kinetic experience as one sweeps through it." Lawrence Halprin, the landscape architect who created a park overlooking a superhighway in Seattle, called freeways "a new form of urban sculpture for motion," and wrote in *Freeways* that "these vast beautiful works of engineering speak to us in the language of a new scale, a new attitude in which high-speed motion and the qualities of change are not mere abstract conceptions but a vital part of our everyday experiences." Kevin Lynch looked to the freeway as a sort of experimental sculpture, whose succession of vistas and objects created a potential for "melodic" design.

Perhaps the surest proof of the sculptural nature of the superhighway is the failure of several desultory efforts to decorate it with sculpture. Even large abstract pieces prominently located become mere details on an Interstate, and appear irrelevant and out of place.

Superhighway construction took much more land than traditional highway building—some 45 acres per mile of Interstate, on the average, and anywhere from 50 to 150 acres per interchange. Most of this, of course, was not highway at all, but shoulder, bank, median, and interchange. Engineers liked to boast about their landscaping of the superhighways as "the nation's largest park"; and in many places the landscaping was lovely, with groupings of trees and bushes, vast extents of grass or ground cover. Some states and highway authorities began systematic programs to disperse wildflower seeds along their thoroughfares. Elsewhere, there were awkward and isolated bushes, like the beautifying boxwoods hastily added by a developer.

These miscellaneous spaces were odd and somehow disquieting. Many of them, cut off in the middle of a cloverleaf loop or set behind a culvert, were rarely trodden by the human foot. To consider them close up makes the odologist feel like a slightly foolish intruder who has lost his way. The sudden switch from the moving, distant, detailless scale in which they appear from the road to close up gives one a sensation of seeing them through a magnifying glass, with every bit of gravel, every bunch of grass distinct.

Some of these miscellaneous spaces could be strikingly beautiful. An intersection on the Interstates might open up into a vista of bunched willows and a gentle bank of grasses, or the median, widening as the roadbeds diverged to meet a hill, would turn suddenly into a miniature meadow.

Even the signs, the reassuring, standardized green, with their cool, metallic supports, silver or naturally rusted brown, played like modern sculpture through the distances and frame of ramp and bridge: confuting plane and solid like some environmental elaboration of David Smith or Anthony Caro.

Other spaces, of course, were wastelands, paved areas at the edge or in the center of the city, jetsammed with litter, walled in with graffitied concrete, monumented by the stripped and burnt-out hulk of a car—an abandoned, stolen, getaway vehicle?—and tinged with urine whose scent was strong enough to invade the passing car.

There was something almost sinister about these spaces where neither car nor pedestrian commonly ventured. Perhaps the most famous of the miscellaneous spaces is the "grassy knoll" of the intersection in Dallas where President Kennedy was assassinated, the locale of phantom gunmen in a dozen conspiracy theories, the shoulder of grass beside the overpass under which the limousine disappears at the end of the Zapruder film.

The median was the original highway no man's land. Originally, it was little more than an extension of the dashed dividing line, rimmed with curb, filled with lawn, parodying a park. As traffic increased and the ease became apparent with which an out-of-control vehicle could leap this token divider, steel guardrails were added, intimidating barriers that squeezed drivers. The aesthetic concern of the sixties saw widened medians, gently depressed to stop errant vehicles, often with skillful plantings.

Designers with their origins in the parkway, like Gilmore Clarke and Michael Rapuano, had begun to separate the two roadbeds completely, fitting each to the topography as if they were two separate roads and turning the median into a sort of parkspace of widely varying width.

In the city or anywhere that right of way was at a premium, the median narrowed to a barrier, first of steel, then of concrete. The dominant form of barrier became known as the New Jersey barrier, after the state that first adopted it. But its history was more complicated.

It had originally been developed in the fifties as the Isleguard, the creation of a man named H. S. Smith, who had watched a friend burn to death when his car caught fire after striking an older steel barrier.

The Isleguard had a wide bottom that diverted the wheels of a vehicle before its body struck the thinner top, preventing crashes and explosions. Its benefits were immediately obvious. But Smith was no engineer; he was an amateur, and the highway professionals resisted the use of his invention for nearly two decades. The Isleguard was not only safer, it was cheaper. Sheer professional obstinacy delayed its adoption until the seventies showed the new power of the highway engineer.

The generation of highway engineers who built the Interstates were different from the traditionalist builders of the parkways. Parkway building, mostly under federal sponsorship, continued to offer a showcase of the vanguard in American highway building art, but most of the action was elsewhere, in the city expressways.

A sort of last hurrah for the parkway builders was the Linn Cove viaduct, part of the Blue Ridge Parkway in North Carolina, in whose construction as little as possible was done to damage the terrain of rocks and creeks. The piles for the viaduct, their shapes suggesting modern Swiss or German inspiration, were precast, trucked in, and set in place with as little damage as possible to the mountainside—a virtuoso demonstration of the engineers' ability to construct highways in a way that would satisfy the most fanatic of their environmentalist opponents.

The old breed had admired the designers of the early parkways. Men like Gilmore Clarke, who worked on the Westchester and Long Island parkways, or Stanley Abbott, who helped create the Blue Ridge Parkway, were landscape architects before they were highway engineers. They laid out their roads by "walking the line," physically traversing the terrain, as Olmsted had in laying out his parks. A professional could immediately recognize the hand of a Clarke or Abbott in photographs of highways they had designed.

But such designers were soon to be extinct. They were replaced by pure engineers, state designers who followed the red and blue books sent out by Washington. With increasing costs and the arrival of higher technology, highway design became more standardized. Highways were designed from maps and aerial photographs, not from the ground.

The victory of the horizontal. The evolution of the bridge overpass, from a Prospect Park carriage road, by Olmsted and Vaux (opposite), to the Bronx River Parkway (right), to a wide, low structure in Wyoming (above). (WYOMING PHOTO COURTESY WYOMING DEPARTMENT OF HIGHWAYS)

9: The Triumph
of the Engineers

In the twenties, when Francis Turner was growing up near Dallas, in farm country that would one day be turned into vast suburban developments linked by superhighways, the state of Texas required each taxpayer to spend time working on the roads. You got $5 a day—the day ran from dawn to dusk—and a dollar-and-a-half more if you brought a team to help grade and ditch the back country roads. It was the oldest form of highway tax.

Turner would often join his father, a railroad engineer, and his grandfather, a professor at a small college, working along the highways to meet the requirement. When it got hot and the men moved slowly and the soil slipped off the drag in a cloud of fine dust, Turner's grandfather would say, "You know, boy, highways are the coming thing. The railroads have just about peaked. If I were a young man I would get into highways."

It was, perhaps, a gibe at the father, but he had to agree. So Frank Turner went to a junior college and then to Texas A&M to study civil engineering, joined the Bureau of Public Roads, and went on to become director of the Federal Highways Administration.

Frank Turner is considered by his colleagues in the highway engineering business as "the man behind the Interstate highway system." Turner reveled in the organizational. "That was a good committee," he would say. He believed in highway design as a triumph of rationalism. He talked about "trips," the basic unit of highway professional thought, and referred to the driver as "the customer."

He was also a paragon of those engineers whose fraternity lay behind the building of the superhighways. Engineers have always been an elite in this country. From the first days of the Army Corps of Engineers, created to fortify American ports in the imminence of the War of 1812 and trained in classic French techniques of fortification, to the engineers of aircraft and space design, hired literally "by the acre" to push some crash bomber or missile program to completion, they have constituted an essential American type. They have been heralded for great, visible achievements, usually after spending the years creating those achievements in obscurity, and have also taken the brunt of the anti-technological waves that surge up regularly from the heart of American pastoralism.

The highway engineers were some of the first true technocrats. They shared a sense of the historical value of their technology. They believed in the ability of research to solve the problems of the highways, and in the national commitment to highways as a high cause. They were men of great probity, dedicated, above all, to the highway "system," sometimes to the point of peripheral blindness.

Turner looks like a kinder version of Vince Lombardi, with efficient horn and metal glasses. All the Bureau of Public Roads engineers seemed to wear these glasses, no-nonsense pince-nez or metal frames, evolving over the years and administrations, but always glasses as an engineer might design them, not the round geometrical glasses of the architect or the "contemporary" style of a fashion designer.

The beginning of road technocracy can be dated back to 1905, when Logan Waller Page became director of the Office of Public Roads. Page was a geologist who had been head of the road materials laboratory at Harvard and had studied at the famous French Laboratory of Bridges and Roads, which had helped make French roads the best in the world.

Under Page, and for years afterward, the activity of the Office of Public Roads was very similar to that of the Department of Agriculture's extension survice: propagandizing rural communities on the usefulness of good roads and the techniques for building them. It sent "good roads" trains around the country to build "object lesson" roads aimed at inspiring local initiative. In this way, it helped build up the rural movement for better roads.

Some of these were constructed using exotic techniques. In areas of the South where clay "gumbo" soil was present, engineers experimented with burning charcoal over the roads to harder the clay into a sort of impromptu terra-cotta road surface. Along the coasts, ground shell was used. On dusty midwestern roads, the first experiments with oil and tar products were made—asphalt pavement had been used in the cities since the 1870s, but was still considered too expensive for rural roads.

The highway engineers spoke of themselves as "we"—meaning the profession, not the state highway department or the federal Bureau of Roads. They were mostly men from small towns, many of them created by the railroad, who saw good roads as a way to improve life in such towns and the roadbuilding profession as a way to get themselves out of them.

Back home, they had been the guys left out of playground and cow pasture games. They took pride in the dull factuality, jargon, and caution of their speeches.

The ultimate highway man was Thomas McDonald, "the Chief," a man with a countenance like the stern president of a land grant college of the sort most highway engineers had attended, who wrote his engineering thesis on the behavior of roads under wagons of

different loading, and who saw historical purpose in the accom-
plishment of the mundane, raised to huge scale. He was a man, a
colleague later said, "of few intimates, who found solace in Bertrand
Russell, Dr. Toynbee, Earle Stanley Gardner, and A. Conan Doyle,
with few writers in between." Wrote another:

> His public speeches ... were so heavily factual and dry as to impose
> a serious burden upon the attentiveness of the audience ... [and] he
> often turned from his written text to interpolate remarks which
> seemed unnecessary and which usually caused both the speaker and
> the audience to lose the thread of what he had been saying. But
> when all this has been said the fact remains that he somehow left
> his hearers ... with a sense of his mastery of whatever the matter
> was that was under discussion.

There was Herbert Fairbank, who in meetings kept twisting a
rubber eraser back and forth until it broke to emphasize his pet
concern for the importance of "dynamic loading," the wear and
tear exerted by moving vehicles. Or the owlish Prevost Hubbard,
champion of bituminous paving specifications, the creator of Medium Curing Cutback Asphalt, the surface that paved thousands
of miles of rural road.

These engineers were singularly upright men. The road program
was free of the sort of corruption that attended the building of the
transcontinental railroad. There were plenty of rigged bids, political
favors, and built-in corporate subsidies, but because of the administrative structure of the highway system—it was perhaps an accidental benefit of the federal/state conflict—there was no Interstate
equivalent of the great *Crédit Mobilier* railroad scandal.

This was due not to the honesty of politicians but to that of the
engineers and technocrats. The location of Interstate exchanges
created one of the most dramatic real estate booms of the century,
with many acres rising twenty or twenty-five times their original
value. Because most of the right of way was brand new—not on
the track of any previous road—fallow fields could become shopping
centers virtually overnight. It was a masterpiece in the repertoire
of that great American form of hysterical theater called real estate.

Once, in a southern state, Turner was offered a large chunk of stock in what was to become one of the largest motel chains in return for disclosing the locations of the interchanges on a planned new highway. "Had I accepted the stock, and I could have, I would be a millionaire today."

Turner went straight from Texas A&M into a special training program the BPR had set up for its federal engineers. He became a pioneer in time and motion studies of highway construction. Industry in the twenties and thirties was obsessed with improving efficiency through such research—originally the creation of Frederick Taylor—and Turner gained a lifelong fascination with the power of research and organization in bringing Taylorism to road crews that had previously been run with no more scientific planning than the Texas road gangs Turner had worked with in his youth.

Turner was dedicated to systems, to studies, to the dominance of the organizational grid. His master's thesis related maintenance requirements to varying grades and subsoils. He came on the scene at the beginning of the golden age of road research, when engineers were developing such devices as roughometers and profilometers, Benckelman deflection beams and Goldbeck pressure cells, electric eyes and tubular traffic counters, moisture gauges, electrical resistivity tests, soil samplers, plasticity indices for soils, the Thin Film Oven Test for asphalt pavements, infrared spectroscopy, tests on load transfer dowels in concrete pavement, vegetation management, and stereo aerial photography for detailed information on terrains.

Some of the road tests were huge, like the $27 million one at Ottawa, Illinois, in which seven miles of pavement and four of eleven test bridges were "tested to destruction," beaten to death by the continuous traffic of huge semis driven by Army draftees, years of wear telescoped together as if in a time-lapse film, first the little cracks, then the bits of pavement falling at the edges, and finally its gradual breaking up.

The results of all this research were boiled down into a Bible of policy and specification books: the *Highway Capacity Manual*, the *Manual on Street Traffic Signs, Signals, and Markings*, the *Manual on*

Uniform Traffic Control Devices for Streets and Highways, Standard Methods of Specifications and Tests for Highway Materials, and finally the red and blue books, as the books of standards for urban and rural roads were termed.

There were also studies of the behavior, capacities, and desires of the typical driver, the democratic atom of the highways, the maps of the "desire lines" derived from interviews. Using these maps and projecting their patterns forward, the engineers laid out the road maps of the future. It was a critical process for highways all over the country: superhighways were to be laid out not according to any larger plan of where traffic and development *should* go, not with any realization of the effects of the highways, but according to where drivers, decades in advance of construction, wanted to go.

The engineers simply projected the desires of the present into the future. In the early forties, for instance, federal engineers estimated that in 1960 Americans would drive 130 billion miles, or half again as many; in fact, they were driving at least three times that figure. They estimated that there would be twenty percent more vehicles on the road in 1960; in reality there were nearly twice as many. The results were congestion, traffic jams, longer travel times—a narrowing of the open road.

Turner loved these studies and models, the advanced science of the profession, but for many years he was sent to build roads in some of the most primitive areas of the world. On the Alcan Highway, tying the United States to its most distant outposts in Alaska, he plotted the route from small planes and served as "expediter." After the war he supervised the rebuilding of the Philippine road system, cutting through red tape to obtain good engineers, and fitting out an old infantry landing craft in which his engineers roamed the archipelago, as the seagoing equivalent of the "good roads trains" he had seen in the U.S. in the twenties.

He returned to Washington to become involved in the Bureau of Public Roads's almost obsessive series of studies for the Interstate system during the forties, years when the Bureau was planning and studying and waiting for the politicians to give them the go-ahead. "We took off with those studies like a guy with a shiny new Mercedes," he recalled. "We really went to town with them."

The studies divided the country into gridded boxes according to income, social characteristics, and other factors, and charted the frequency of trips from box to box.

"In each box," Turner later recalled, "we conducted sample surveys, like the political polls or the census bureau, with intensive home interviews of a statistical sample. The percentage of error was four percent, plus or minus."

As Federal Highway Administrator, Turner caught the brunt of the opposition to highways that developed in the late sixties and early seventies. Like many of his fellows, he developed a sort of war-weary tone, a sense of public incomprehension and lack of appreciation for his work, a scorn for what he regarded as the fashions of the times. "We were environmentalists from way back," he would say.

Turner presided over the highway program at a time when thousands of people were displaced from their homes and thousands of small businesses destroyed to make way for the Interstates. He expanded the compensation program for those displaced. "We had to buy out lock, stock, and barrel virtually all businesses where we took away their right of access if we could not restore it some way with a frontage road, a secondary access road."

As the top highway boss, Turner would travel regularly to places where a recalcitrant owner refused to move: a motel along Route 66 being replaced with Interstate 40, or a series of homeowners threatened with an urban freeway.

Turner's own parents were displaced, in fact, by I-35 near Fort Worth. "They said, 'Son, can't you do anything?'" he recalled wistfully some years later. "And of course I couldn't. Well, they didn't want to leave the old place, but when they got settled into the new house, all brick, they were happy."

Turner seemed to feel only a limited satisfaction at his rewards, national and international; a bit more perhaps, from the photograph in his basement showing the Clay Committee—of which he was executive secretary—presenting the Interstate report to President Eisenhower.

What really counted for him, in his retirement, was the respect of the fraternity. AASHO would call him out of retirement to address its annual conventions. Amid the workshops on such current topics as noise suppression walls and wildflower plantings, signs of the profession's need to placate a public once so enthusiastic about its work, Turner would recall the grand vision of the national highway system.

He might discuss his scheme, never adopted, to standardize the designation of Interstate interchanges, which had varied from state to state, with a system based on mileposts.

"I wanted to make the mileposts bigger," he said, "and number the interchanges according to the mile: I-95 at mile 200. That way you would also know from the mileposts just how far you were from the next interchange.

"With this system we could have given a number to everything on the system—every sign, culvert, and so on. We could have put it in a computer tied to our accounting system: a ten-digit number that would let us know just what we had spent and what we had repaired and the age of everything on the whole system.

"Eventually, it would replace Zip Codes. Everybody would have an address that would be one of these numbers."

Turner's computerized system was never adopted. But by the eighties, the engineering knowledge of many years had been shaped into computer programs. Stereo aerial photographs were digitized and fed into the system. $50,000 Intergraph electronic drawing boards with millions of "addressable points of resolution" projected potential highways from the driver's eye. The planner could leap forward a hundred feet with the touch of a button, curve around a hill, then back up and try an alternative route through its edge. He could shift from a steep cut to a shallow one, from a banked edge to one protected by a standardized guardrail, stored in the computer's memory, then zoom closer, even picking out the nuts and bolts on guardrails. The same Texas Highway Department that had once relied on amateur labor now boasted one of the most sophisticated of these computer systems.

Howard Johnson's, Queens, New York. An exaggerated Mount Vernon, the flagship of the Howard Johnson's fleet, 1940. In the background to the left, the Trylon and Perisphere of the 1939 New York World's Fair. (FENNO JACOBS © 1940 TIME INC.)

THREE
The All-American Roadside

Strip Ahoy, St. Petersburg, Florida. The essence of stripstream architecture: a shipshape striptease joint, with riverboat smokestacks, portholes, and long, smooth shape.

10: *The Great Striptease, or An Ecology of the Asphalt Jungle*

"Be prepared to see more than you expect."

—*Motto of Roadside America, Shartlesville, Pennsylvania*

Larry Gieringer, with his wife Dora, built up Roadside America, "the world's greatest indoor miniature village," along Route 22 in Pennsylvania. The motto Gieringer painted near the hex signs on the vaguely Pennsylvania Dutch facade of his building falls all over itself, like Dr. Johnson's line about the elephant who claimed, "I am the largest elephant in the world—except for myself."

Gieringer's advertising line could stand as a motto for the whole curious ecology of roadside culture that is the American strip. The strip is made up of enterprises that all want to be reassuringly familiar, as "expected," with "no surprises," as in the Holiday Inn ad line, but also different, special, unique.

The strip is a realm of boast and promise, an ecology of advertising, an informational jungle where only the strongest images survive; where everything claims, by a thousand subtle signs and designs, to be fancier, bigger, cleaner, faster, better than it appears to be, where the quaint Main Street values the strip has replaced are recapitulated in caricature; where the skill of all promises is to inspire a credulity bordering on the suspension of disbelief in the most incredulous of passersby; and where winners and losers, survivors and casualties, are distinguished by the degrees of their success in luring drivers off the road.

There are standardized witty plaques on the wall behind the grill at Roadside America—"Old fishermen never die, they just smell that way" or "The opinions expressed by the husband of this household are not necessarily those of the management"—and shelves full of Pennsylvania Dutch souvenirs. On the wall is posted the legend of how, in June 1903, at age nine, Larry Gieringer went up to Mount Penn above his boyhood home of Reading, looked down upon the town, and came back down, like some barefoot, juvenile Moses, with the calling to build Roadside America.

This model village is not much, but, having been urged to it by signs along the way—many fewer now than once—you are eager to visit. Inside the darkened space you encounter a sort of gigantic model railroad set, representing almost every aspect of old-fashioned bucolic and small-town life: the railroad depot, the general store, the church, the quiet residential street.

There is a working country barn dance with moving figures, log cabins and blacksmith shop, moving figures cutting wood and pumping water—6,000 gallons of water an hour flow through the streams and rivers of the display. There is a model of the home and store in which Larry Gieringer was born, the son of a harness maker. A village of Indian teepees stands beside a coal-fired electrical generating plant. Scattered throughout are ninety-six horse-drawn carriages, complete to the last detail of their harness work.

To the odologist, Roadside America suggests nothing so much as a parody of Norman Bel Geddes's dioramas of America in 1960: this is the oldtime America Bel Geddes wanted to streamline away, the small-town Main Street America of Disneyland or Henry Ford's Greenfield Village, celebrating the "turn of the century" America that the automobiles killed. There are no roadside attractions in the model at Roadside America.

Roadside America is one of thousands of old-fashioned, do-it-yourself attractions of the highway—the entrepreneurial alligator and snake farms, the ghost towns where prerecorded gunfights take place, the Appalachian rock shops and the Florida shell shops, all aimed at luring the vacationing driver and his family off the road. And like most of them, it aims to conjure up the world as it was before the road, the frontier world, the historical world, the natural world. The road that draws tourists has always been a road of nostalgia.

The arrival of the automobile almost immediately brought the first roadside attractions onto our highways. Any place that was good enough to locate one restaurant or gas station was good enough to locate two or three or more, and the strip was born. At first the strip was an extension of the tourist attraction, radiating from beach or resort. But as time went on it became a feature of the everyday roadside, a bit of entertainment along the dull route of life, a touch of permanent vacation.

The ugliness of the strip was already lamented in the twenties and thirties. But its architecture was also recognized as powerful, creative, and, somehow, completely new, with a lure all its own.

The strip, its critics say, is the same everywhere. This is partly true, and part of its appeal. But the strip also contains "local color," neatly reduced to a few handy icons that work—in the tourist strips: the nets and lobsters on the walls of the seafood restaurants in Maine, the giant muskie in the fishing country of Minnesota, the mock L'il Abner cabin and still in Appalachia, the alligator farm selling alligator wallets and key cases in Florida, the café with mock Navaho sandpaintings on the walls in Arizona, the souvenir stand near the Petrified Forest, with pen and pencil holders, paperweights, and wall clocks whose Gothic hands trace out the hours on a face of the sliced stone that was once wood.

The ugliness of the strip is due less to the buildings themselves than to visual static: the conflict of styles, the intrusion of power and telephone lines, all the visual white noise. These distractions fight the flow of the road; they assault the driver with an exhausting dissonance of messages.

The ugliness of the strip lies in its miscellany, its clashes of style and message, its provincialism and cheapness. It is a garden mostly of weeds, but with the vital force of weeds, a field in which, from time to time, beautiful wildflowers appear.

But from the beginning there was positive feeling for the strip and its architecture, and sometimes from surprising quarters. In 1936, the architecture critic H.-R. Hitchcock, who with Philip Johnson championed the International Style in a book and an exhibition at the Museum of Modern Art, and who was a fan of Frank Lloyd Wright, injected into a book about the architect H. H. Richardson this startling estimate: "The combination of strict functionalism and bold symbolism in the best roadside stands provides, perhaps, the most encouraging sign for the architecture of the mid-twentieth century."

Hitchcock referred in particular to the Freda Farms ice cream stand on the Berlin Turnpike in Connecticut, whose building was shaped like oversized cardboard ice cream containers, arranged like a Palladian villa. A large container formed the center of the store; two wings of smaller containers spread out to either side like the service facilities of the Villa Barbaro. Mounds of pink "ice cream" and jutting "spoons" topped off the roof.

The roadside strip began as the province of small business, of do-it-yourself design. Even though dominated today by the big franchises, it still assumes that air. It is a fair model of uncontrolled free enterprise.

The design of buildings on the strip is based first of all on survival by distinction: it is a kind of anti-camouflage, a struggle to stand out. The architecture of the strip was the first where words reigned supreme over decoration and signage replaced ornament. The giant burger rotating atop a steel pole is a successor to the boot in front of the cobbler's shop, the painted wooden fish in front of the fish-monger's. Those images were there for people who had never learned to read. The modern images are there for people moving too fast to read. Their aim is to catch the distracted eye.

What makes Larry Gieringer's model work is the care he took to create everything in exactly the same scale throughout the model; what makes it impressive is the fineness of the detailing.

On the real roadside, nothing is in scale. Everything is larger than life. Scale on the roadside is an advertising medium, a device for eyecatching visual jokes and "tall tales." And as long as the shape and impression of the roadside world is right, its details are unimportant.

There is a principle of natural selection that, for different reasons, also applies to the roadside environment. The physical size of a species, this principle holds, tends to increase with each generation. The larger the specimen, the better chance it has to reproduce itself—up to the optimal size. This is also true on the strip, where structures must be perceived over distance.

The artist Robert Morris once created a piece in which he employed the services of a woman blind from birth to draw while he spoke instructions to her. At one point in the execution of the piece, he tried to explain perspective to her. "Objects further away appear smaller than closer ones," he said.

"That," she replied, "is the most ridiculous thing I've ever heard."

Perspective and the scale of distance remain concepts that are difficult to rationalize. We accept them through experience, not intellectualization. We know how things look across distance, but not why. The laws of perspective are as fundamentally elusive as those of relativity. If we spent as much time traveling in spaceships close to the speed of light as we do walking or driving, we might well take relativity as much for granted as we do perspective.

The road brings back some of our primitive wonder at perspective—and our sense of its humorousness, the absurdity the blind woman referred to. The very course of the road is marked by a narrowing recession. The speed of the observer moving along it abbreviates and telescopes the experience of perspective. It also adds other factors: the psychological narrowing of the field of view—we see much more ahead of and therefore distant from us in the car than along the near sides—and a more rapid succession of what perceptual psychologists call "sight fixations."

The do-it-yourself builders of the first roadside enterprises knew all these things, mostly unconsciously, and the most sophisticated creators of the later franchise and chain structures learned them consciously. It was no wonder that buildings along the roadside quickly grew in size or that their architecture came to consist of icons and visual puns.

The wonderful giant donuts, the tea and coffee pots, the big pigs and ducks, the camera-shaped film shops and beer halls in great kegs moved from the humorous to the surrealist, like those classic postcards of fantastic animals: jackalope, rideable jackrabbit, trout leaping over the boat, potatoes and apples on flatcars—surrealist like Magritte's apples, but created by spurious collage. They were as Pop as Claes Oldenburg's soft fans and vaccuum cleaners, his giant clothespins and baseball bats.

Given the perspective of the road, the giant donut, first appearing over the horizon of the dashboard, might strike the eye as about the size of a real donut: a pun of perspective.

Since the real structures in which the early roadside businesses were carried on were little more than stands, often homemade, the sign was by far the largest part of the design. Why not then make it one and the same structure? Why not make the building the sign?

Of all the Main Street values the strip embodies, none is stronger than that of the do-it-yourself inventor. And the inventors of the most colorful strip structures all saw themselves as entrepreneurs. No one aimed to build just one giant ice cream cone or cowboy-hat-shaped gas station; the idea was to develop a chain of tens of the things, to expand on the invention. The patent office in the twenties and thirties was deluged with drawings of dirigible-shaped sandwich stands and oil-can-shaped gas stations. William Altman of San Antonio, Texas, for instance, received U.S. Patent number 90303 for his pig barbecue, its details specified right down to the screen door in the creature's belly and the pattern of painted spots on its back. But only a handful of the little stands were ever built.

Carried further, the tradition of gigantism became more indirect, especially if it better served the function of the enterprise. A Mexican restaurant, for instance, might display not a giant tamale (as others had done) but a giant sombrero, which carried the message more abstractly, and in the process provided a convenient overhang to shelter customers and even, in one Tijuana example, a balcony from which visitors could survey the surrounding landscape.

The need for differentiation also led to more indirect symbols. When there were two or three ice cream stands with giant cones for signs, the creative owner of a new stand might resort to an abstract sort of expression: the stand shaped like an iceberg or like an ice cream freezer, complete with handle. The next man might have to go to a still more complicated visual/verbal pun: a giant owl, with the vending window in its breast and the explanatory sign: "Hoot hoot, I scream."

There were other traditions for the roadside entrepreneur to turn to. Architectural historians have found precedents in the late eighteenth century French "rationalist" structures of Etienne-Louis Boulee and Claude Nicolas Ledoux, like the latter's house for a woodcutter designed in the shape of a pile of wood. The similarity is real, but it is not likely that they were sources for ice cream magnates or hot dog kings.

More to the point is the whole American tradition of gigantism, from the tall tales concerning Davy Crockett or Paul Bunyan, via the whole ethic of boast and exaggeration—abetted by willing audiences.

Some of the best-known roadside giants, like the famed Brown Derby on Sunset Boulevard, or the huge cowboy hat and boots of a Washington state gas station (the boots housed the rest rooms), recalled a tall tale whose point was the horrendous condition of American roads. A traveler came upon a beaver hat in a huge puddle in the middle of the road. But when he reached to pick it up, a hand emerged from the puddle and he heard a submarine voice, "Whoa there, brother, Old Nellie just found the bottom." Roadside gigantism was a practical joke. It was a flippant bet that led Herbert Somborn to build his famous Brown Derby restaurant in Los Angeles. "If you serve good food, then people will even eat out of a hat." Legendary story or not, the principle was a key one for the roadside.

The great elephants constructed in Coney Island and, even earlier, in the seaside resort of Margate, New Jersey, are the first known examples of gigantic entertainment architecture. They helped tie roadside architecture to an earlier tradition, that of P. T. Barnum and his famous elephant Jumbo. The roadside was interpreted as a circus or carnival in which the audience, not the acts, did the moving.

Amplified by the perspective of the road, oversized scale became one of the roadside's most effective illusions: Is it big or is it just closer than it looks? The driver must be prepared to see larger things than he expects.

The modern auto strip also has in its lineage another sort of strip: the western towns rapidly built up along the railroad, like the Cherokee strip of Oklahoma, a block of land that was opened up to settlers at noon one day in 1889 and twenty-four hours later boasted a "town" of 10,000 people. It was built fast, of tents, prefabricated buildings, and "box" buildings—cruder versions of balloon frames. Its main street boasted false fronts, and behind many of the fronts was nothing more than a large tent. It was one of the last of the fast western towns, and it may have been the fastest.

Some of the earlier western towns literally arrived on flatcars. They were called "Hells on Wheels," because they brought with them all manner of gamblers, camp followers, prostitutes, and miscellaneous bad guys. They were the wild towns the marshals and sheriffs tamed in classic Hollywood Westerns.

The Old West was a stage set, a board and canvas simulacrum

of the Old East. Aspiring to the established feel of towns back home in Ohio or Connecticut, western towns were ghost towns as soon as they were built, haunted by intimations—perhaps one should say intimidations—of the East.

The basic unit of development along the railroad was the railhead town, a cluster of buildings near each station. The motor road, however, did not restrict development to the nodes of stops. What had been concentrated by the railroad into a few facades on Main or Depot streets, was spread by the road along its leisurely length. Crossroads might be important, but land prices and the need for direct frontage naturally led to the horizontal extension that came to be called the strip.

In Los Angeles, in particular, the western tradition was brought to a head by the world of the movie backlots—in the form of wild Hollywood stage-set restaurants in the guise of Dutch windmills and farmsteads, or colonial mansions (like the Plantation Inn in Culver City, owned by the infamous Fatty Arbuckle).

Perhaps the ultimate roadside attraction as western town is Old Tucson, near the Arizona capital. There are tourist cowboy towns all over the country, from New York and Tennessee to Colorado and Wyoming. Most were built from scratch, a few are reconstructions of old ghost towns. But Old Tucson is different. It is at once a movie set where such films as *Arizona* and *The Wild Bunch* were shot and a tourist attraction, with staged gunfights where actors play to recorded sound effects blasted through speakers. Old Tucson is attractive, not because it is a genuine town, but because it is a mock town, the force of whose artifice is confirmed by its success in the movies. It is a place, like so many on the strip, where people come half to be deceived and half to marvel at the cleverness of the means of deception.

The ship is a frequent form for roadside "vernacular" architecture. The road, after all, was a kind of river—a stripstream of construction and imagery—and the whole system of roads, with the land cruises they made possible, not unlike a sea.

When the building in question was a seafood restaurant or located in a tourist area near the ocean the analogy became even more convenient. One enters many such buildings via a gangplank. Often there are telephone poles driven into the ground, as if for moorings, where you fairly expect pelicans to settle.

Roadside shipshapes run from chain outlets—California's Chicken Supreme or the Southern Kingfish chain—to multipurpose buildings like the S.S. Flagship on Route 22 in New Jersey. This ship-shaped structure, once a dance hall, became a discount furniture outlet. It is a mildly famous building, having been featured in two or three of the slew of roadside nostalgia books that came out in the seventies, but its "gunwales"—actually rickety fencing material—have fallen down, and its shape is marred with so many miscellaneous airconditioning units and other additions that it barely strikes the inattentive viewer as a ship at all.

More famous is Robert Derrah's 1936 Coca-Cola Bottling Plant for Los Angeles, with rounded corners, portholes, a superstructure with a captain's bridge bearing Coke's logo, and, on the "stern," an oversized bottle.

The stripstream is an appropriate place for streamlined architecture—whether the "natural" streamlining of the boat shape or the analogical streamlining of the Moderne, picking up on the ocean liner shapes that twentieth century architects have always loved.

Roadside vernacular was never far from streamlining. The classic, shiny, metal diner imitated the sleek, silver railroad cars of the Twentieth Century Limited so successfully that the average customer viewed it as a dining car taken off the tracks. In fact, the diner was an evolution of the simple lunch wagon, now brought up to date by streamlining.

Streamlining was Moderne, not modern. Modern was functionalism, or expressionism disguised as such. Moderne was a decorative idea of what the architecture of the modern age, the age of the machine, should look like. It grew from a widespread sense, or wish, that we were already practically living in the future, and the future was the road.

Another nautical shape, the riverboat—with its generally square shape, shallow draft, and lack of the streamlining of faster ships—particularly lent itself to dry-land adaptation. It was itself a piece of architecture—at its height, the riverboat tried to be a grand, if boisterous, floating hotel and music hall—a showboat. A Mc-Donald's near the Gateway Arch in St. Louis is a riverboat-style structure—estimated to have cost a million dollars—that actually floats. Back on land, the riverboat provides an excuse for ginger-breaded balconies and chandeliered dining rooms.

The boat on land is a showboat whose associations extend to yachts, with overtones of rich men's parties—perhaps hosted by a bootlegger like Gatsby, or the gangster of *Farewell My Lovely*, operating an offshore casino.

Perhaps the ultimate ship/strip tease, called Strip Ahoy, stands in the long strip of U.S. 19 running through St. Petersburg, Florida. Once it was a nightclub. Now its curved walls and tempting but darkened portholes enclose a club dedicated to what its owner good-naturedly calls "an all-female revue." Inside, the boat imagery fades fairly quickly into a conventional bar and burlesque joint. The name says it all: the grounded ship, like all essential strip architecture, is a put-on that slowly, coyly, teasingly removes its pretensions.

Strip Ahoy is a token of the fact that, on the strip, the boastfulness of illusion and allusion are almost always deflated by some glaring variation: the dumpster visible behind the place, the undisguised back of a giant sign. This is all in good fun; it is roadside theater.

The stripstream was the descendant of the river, and on the river both Melville's Confidence Man and the King and Duke of *Huckleberry Finn* are at home. There is a will to self-delusion here, an almost theatrical suspension of disbelief. The spectator is actively involved in creating the fiction of existence, a co-conspirator in his own befuddlement.

At the heart of the experience of the strip is a characteristically American epistemology: with things flowing, with an awareness of perception as a process, there was no need to look too closely for the essence of objects. Nowhere was America's radical empiricism more at home than on the strip: it looks tropical, it looks French, it looks colonial, so it must be. Perception was subjective; it involved a will to believe. Things were what they looked like.

The strip was the descendant not only of P. T. Barnum and his conviction that the public liked to be deluded for its own entertainment, but also of William James, who argued that perception was less like a "train" of consciousness than a "swarming continuum."

Traditional empiricism, with its disturbing implication that there was really nothing at all out there, became much more palatable when the element of change, of flow, was added.

The rulers of the stripsteam are like Twain's King and Duke: con men of architecture, with their easily debunkable pretensions to being castles, chateaus, or mansions. As the King says, apropos the bogus Shakespearean drama he puts on under the title of *The Royal Nonesuch*, "it's the histrionic muse."

The little river towns where the King and the Duke mount their shows are bored, dull towns, Main Street towns, whose citizens are so eager for novelty that they will accept being fooled for it, and whose architecture is poor and provisional. Their houses and stores are "mostly all old shackly dried-up frame concerns," "all . . . along one street" and constantly eaten away by the erosion of the river. "Sometimes a strip of land as wide as a house caves in at a time. . . . Such a town as that has to be always moving back, and back, and back, because the river's always gnawing at it."

"What was the use to tell Jim these weren't real kings and dukes?" says Huck. "It wouldn't a done no good; and besides, it was just as I said; you couldn't tell them from the real kind." The same is true for the average driver: the architecture roadside buildings imitate strikes the average American as no more honest or genuine than its replicas.

On the strip, the elements of this historical architecture are simple and abundant. There are chateau steak houses, a seafood place with a New England widow's walk, a mock Chinese temple, a disco shaped like a thirties vision of a spaceship.

Yet some historical elements of roadside architecture are, at least in outward form, genuine survivals of earlier inns, courthouses, or churches. The cupola is one of the oldest historic elements of road-side architecture. The cupola suggested Independence Hall and Mount Vernon and old courthouses. It was used by Howard Johnson and A&P supermarkets, by 7-11 convenience stores and Atlantic-Richfield gas stations. It was a survival from the earliest days of the road. In the twenties and thirties, America was colonial-mad, feeling, perhaps, a reaction to all the novelty of the post-World War I years, and a nostalgia for the last time when it was securely anchored to an older tradition.

Signs of past times. The broken pediment, an architectural sign of welcome, survives on the modern strip, a descendant of the inn sign and the Georgian entrance.

II: Signs

"A place is only as good as its sign."

—*James M. Cain*, The Postman Always Rings Twice

Very early in its history, the motor road became a realm of signs, where images become icons and words have their own special, free-floating logic—a logic of boastful adjective and typography that overwhelms literal meaning and just as buildings along the roadside evolved into gigantic signs, the signs themselves moved toward the thickness and bulk of buildings.

American signs have long aspired to the third dimension. The great, shaded letters you can still see fading on brick walls in older American cities, the signs on old western storefronts are shadowed, highlighted, and blocked toward solidity. Beginning in the twenties, roadside signs took the dramatics of Times Square as their models. They became shaped boxes of metal, with lights in them. The arrival of relatively inexpensive neon in the early twenties put a touch of Broadway, a bit of Hollywood, at the disposal of the roadside merchant.

The do-it-yourself roadsigns of less sophisticated enterprises resorted to other techniques: clever names and images, intersecting in a mentality whose descendants are visible in the useless pop items of T-shirt parlors, souvenir stands, and boutiques across the country. In these signs, punning, alliterative, and visually catchy names like Dew Drop Inn and Klean Kozy Kabins were joined by barbecue signs featuring complicated rebuses with a bee and a pool cue. A hot dog stand was very quickly turned into a "doggery" or "kennel," and "Ye Olde," as a designation accompanying the popular mock-Tudor style of architecture, was a roadside joke by the end of the twenties. This personal, mobile sort of language had by 1926 produced the lasting term "motel," and by the late thirties the "drive-in."

Gertrude Stein found something exemplary in the language of the road sign. In her lectures on American writing she commented, "... in advertising and in road signs, you will see what I mean, words left alone more and more feel that they are moving and all of it is detached and is detaching anything from anything and in this detaching and in this moving it is being in its way creating its existing."

By the fifties, road signs had to work harder, reaching out and collaring the driver with leaning poles, curving arrows, boomerangs, kidneys, and other biomorphic shapes inflected toward the entrance of the place they advertised.

With the liberal attack on the chaos of the roadside in the sixties and the regulation of billboards, road signs began to grow smaller. Local zoning and other laws led McDonald's to shrink its arches and Kentucky Fried Chicken to remove its giant buckets. The more successful the franchise operation, the smaller the sign needed.

Consider the fate of the Holiday Inn "Great Sign," half firework show and half marquee: "Congratulations Jimmy and Darlene"; "Welcome Plumbing Spec Analysts"; "Happy Hour 4 to 6." The Great Sign was a sort of fifties Moderne warbonnet, designed by one Eddie Bluestein. With its twirling star, its arrow reaching out like a vaudeville cane to pull the motorist in, and its theater-style marquee, it became an American icon.

But in the early eighties, Holiday Inn abandoned the Great Sign. It was an electricity guzzler; neon everywhere had gradually fallen out of favor in place of less dramatic plastic panels, backlit from within. It was also too garish. It retained the lean-out-and-grab-'em attitude of the fifties, an attitude that had been replaced by the more subtle ministrations of television. Holiday Inn's reputation, refined into an "ambiance" by television commercials promising "no surprises" and an electronic reservation system featuring toll-free phone numbers, had replaced it. The decision to stay at a Holiday Inn, more than three-quarters of the time, was made in advance, by reservation.

Similar changes altered the signage of even the smallest roadside operator. He no longer had to paint his own sign. He could rent or buy a standardized signboard, little more than a marquee, with movable letters, and on its top an arrow lined with bulbs—a universal sign that itself arrived on wheels, towed behind a pickup truck.

Another progression characteristic of the iconography of the roadside was the change that took place in McDonald's Golden Arches. From the most prominent architectural element of the chain's stores it moved easily to become the company's sign and logo.

The original 1952 yellow steel arches of the first McDonald's, before Ray Kroc bought out the brothers, held up a tiled back roof above walls of shiny red and white tile, accented with red, green, and white neon. The arch was repeated in the sign out front, a single arch bearing the company name, the words *self-service system* and *hamburgers*, and a space for posting the familiar running count of burgers sold. The forgotten genius behind this store was architect Stanley C. Meston of Fontana, California.

After franchising, the hyperbole of advertising squeezed the arches into steep parabolas. Paired to echo and later to replace the "M" in McDonald's, they were reproduced on napkins, cups, boxes, coffee stirrers, seat backs, garbage containers, and, as stumpy little signs, entrance and exit markers. When billboard restrictions arrived, they ascended to the top of huge steel masts rearing high above superhighway intersections, and as local building codes began to ban large street signs, they were flattened and shrunk. So successful, so universally known had the company become that it was eventually willing, in a particularly resistant Maine town, to build a store without any arches at all. A simple sign bearing only the McDonald's name now held all the visual resonances of the arches in its initial.

The billboard was a familiar fact of life well before the automobile. Innovations in printing made large signs possible by the 1880s; the circuses of Barnum and the Ringlings, significantly, were among the largest customers for such signs. Civic organizations were already complaining about billboards by the 1890s. By the turn of the century the twenty-four-panel billboard had become standardized in shape and size, like the movie wide screen, whose proportions it shared. The billboard was a mass medium.

It managed even to survive the head-on assault of liberal beautification forces in the 1960s. During one committee meeting on the Beautification Act, John Houck, head of the billboard lobby, which for years had boasted of its "self-policing," managed to get a provision tacked onto the bill providing an exemption to the ban of any billboards within 600 feet of Interstate highways.

Signs could not be built closer to the Interstates, the provision said, if the land was zoned for commercial or industrial uses. So thousands of tiny patches so zoned—ostensibly minigarages or parking lots with space for two or three cars—were created simply to serve as sign bases, launching pads for hundred- or hundred-and-fifty-foot signs costing $50,000 or more, with huge standards of steel pipe telescoping out of one another like the tapering stages of a huge rocket booster.

The standardization of official signs began in the mid-twenties, replacing the rainbow of colored telephone pole blazes that marked the old, privately funded roads. Still, a federal route might often also be a state route and a county route. Four or five numerical distinctions might apply to a single stretch of blacktop, and at intersections the motorist would frequently be confronted with huge assemblages of signs, a dozen or more on a single pole, marking north and south, temporary routes, detours, and so on.

The superhighways brought their own system of signage, nationally standardized in color and shape. The reassuring green indicated basic route instructions, blue stood for various services, and historical areas were indicated by a woodsy brown. In place of the massed badges and shields of the past came huge green slabs, hovering over the pavement, arrowed and reflectorized, rationalizing the accretions of routes.

The power of the road sign was confirmed early on by advertisers. The concept of road signs in sequence—the milestones become conceptual—created the memorable Burma Shave jingles, the running joke of Wall Drugs ("Wall Drugs, Wall, South Dakota, 780 miles"). Such a network of signs in effect *created* the "attraction" at its center, as in the cases of Rock City, South of the Border, or, in one wonderful example along Route 66, the Club Cafe in Santa Rosa, New Mexico, featuring the grinning, hard-sell face of an owner who looked like Joe McCarthy or a used-car salesman. Little matter that no attraction could live up to the workings of individual imagination inspired by such signs and fermented over hundreds of miles of travel. The places were manufactured goals of journeys, reference points, and as such they held a value beyond their immediate entertainment. They provided the right to say, "Wish you were here" on a postcard and "I've been there" on a bumper sticker.

South of the Border, off U.S. 1 and then I-95 near Dillon, South Carolina, was an attraction familiar to millions who had never been there, through bumper stickers and punning billboards up and down the east coast. Originally begun as a stand to sell beer to visitors from the neighboring dry state of North Carolina, it grew on the basis of its location about halfway on the New York to Florida tourist route and an advertising campaign based on a geographical pun: just south of the North Carolina border, it pretended to be a little bit of "down Mexico way."

An attraction like South of the Border was the sum of its signs, of the expectation and imagination built up in the minds of tourists beginning hundreds of miles away. Once they arrived, their visions of the place were so fixed that they either saw what they imagined or were too embarrassed to say they were disappointed.

When they reached the place, directed by a sober green Interstate exit sign, the tourist found a hundred-acre sprawl of nothing but a miscellany of tourist services and souvenir shops that could have been anywhere, joined by the relentless Mexican stereotypes: a giant sign of "Pedro," a sombrero-shaped restaurant, a motel, tennis courts, a "campground" for recreation vehicles, "Amigoland Mexi-Mini Golf," and shops offering discount towels, Mexican and Confederate souvenirs, and fireworks. As the ads put it, "Sometheeng for Every Juan!" Something—and nothing.

South of the Border was a pastiche of styles and attractions like the food served in the Sombrero Room—"Confederate Cooking, Yankee

174 Three: The All-American Roadside

style!" or "Pedro's Hot Tamale, ze mos' beeauteeful fast-food Restaurant een ze South!" And its appeal was as broad as its sense of humor. "All roads lead to Pedro's, Amigo, so y'all come, olé?"

Except in such cases and on classic, showcase strips like Hollywood Boulevard, billboards tended to become more discreet. Billboards for Holiday Inn, which had once displayed the image of the Great Sign, became much quieter. Standardized nationally, they featured simple statements of the distance to and location of nearby Inns. There was no pitch, just information. And the information was printed on a background of green almost exactly the same shade as that of signs on the Interstate system—a friendly, familiar, forest green that left no doubt of its message: that Holiday Inn had become as universal as the superhighways themselves, that, with some 600 outlets on their interchanges, it was the closest thing around to the official hostelry of the Interstate Highway System.

The changes reflected a change in the Inns' role. In many towns and small cities, Holiday Inn and its competitors had taken over the function of the old main hotel, hosting conventions of library clubs and county commissioners, the regular meetings of Optimist Clubs and Bar Associations, and even weddings and banquets.

Perhaps the ultimate road sign is a sign that marks the end of the road: the Hollywood sign, where real estate blends with show business. Created in 1923 to tout a real estate development, it became something of a logo for all Los Angeles, of what millions drove cross-country to find, an equivalent of the Statue of Liberty or the Golden Gate Bridge.

Like most signs, it was never intended to be permanent, although for many years a custodian—a sort of permanent prop man—lived in a shack behind one of the "l"s. It was built of steel, guy wires, and paneling, like a movie backdrop, and lit with 4,000 twenty-watt bulbs. It was quickly recognized as a landmark. The sign was donated to the city in 1945; when it began to deteriorate in the 1970s a group of movie stars and executives contributed funds to restore it, with various celebrities, including Gene Autry and Hugh Hefner, each donating $27,700 for a letter.

It was not even really in Hollywood; it bore testimony to the fact that Hollywood was a state of mind and a business, without geographical truth. It was a symbol of fleeting and false hopes, will-of-the-wisp dreams as shoddy as its construction. Sometime in the thirties a frustrated young starlet is supposed to have plunged to her death from its edge, in full view of the motorists below.

What the cupola was for buildings, the pediment, usually broken, was for signs. At least half the signs on an average strip, by one odologist's calculations, feature this architectural element. Some carry it out in wood, some in plastic, some in metal. The International House of Pancakes uses a broken pediment, as does the occasional Computerland.

There are good reasons for this. The pediment top on a sign was a feature of inn signs in the eighteenth century. No one knew quite what a sign was then: was it a part of the architecture of the building, or a sort of exterior piece of furniture? The broken pediment put the sign at peace with both interpretations. It became the sign of a sign.

A midwestern chain called Bob Evans' restaurants has created an apotheosis of the broken pediment. Its homestyle menu is highlighted by bright red buildings done up in Ohio gingerbread ornament, topped with pediment. The sign carries out the gingerbread theme with a sort of Corinthian frieze, attached to a huge broken pediment on top of a steel shaft. Even Caesars Palace uses a version of the pedimented sign.

The pediment suggests Francis Parkman's famous description in *The Oregon Trail* of the old furniture he saw discarded by pioneers along the way—chairs and cupboards that, he imagines, might have come by sea from England a century before and then been carried across the Alleghenies, but which, at some critical juncture on the road, had become too heavy for the wagon and been abandoned, "flung out to scorch and crack upon the hot prairie." Like these, the cupola and pediment are signs of all we have had to discard and all we are trying to remember along the highway.

Do-it-yourself architecture. Pig-shaped barbecue stand, Harlingen, Texas. U.S. patent number 90303 — the model for a nationwide chain that never was. (RUSSELL LEE, LIBRARY OF CONGRESS)

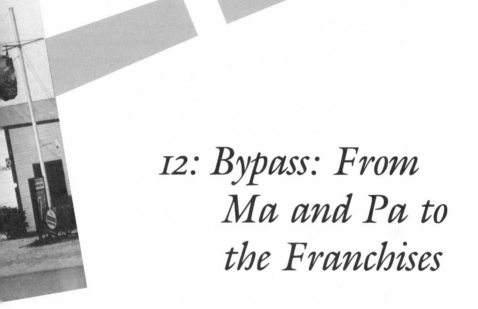

12: Bypass: From Ma and Pa to the Franchises

In the early sixties, the old U.S. Route 22 was bypassed by the arrival of Interstate 78 and Larry Gieringer's Roadside America was left high and dry on a bluff beside the four-lane. Gieringer fought the state highway department and finally forced it to provide him with an access road and a neat green official Interstate sign acknowledging his existence. Most other roadside entrepreneurs were not so lucky. They either moved or went out of business. Entire towns grown used to tourist traffic down their Main Streets faded into obscurity.

Roadside locations with access became premium properties, increasing many times in value as the Interstate and other new highway systems were constructed. In such high-priced locations, only the proven enterprise, usually the franchise, could survive, bunched together in little villages, "easy on, easy off," as the signs boasted, shelved above the superhighway and centered on nothing but the bridge and off-ramp, the cloverleaf their village green.

One town bypassed by the Interstates, similar to and yet different from so many others, was Corbin, Kentucky.

The natural feature near Corbin of which the town is proudest is a moonbow. A moonbow is the nocturnal version of a rainbow: a miraculous deployment of the spectrum through a water spray, with a bright moon instead of the sun supplying the light. There are believed to be only two natural moonbows in the world. One is at the foot of Victoria Falls on the Zambesi River in Zaire. The other is at the foot of Cumberland Falls, in the green and gentle Kentucky hills near Corbin. Corbin, oldtimers will tell you, was created by the railroad, nurtured by neighboring coal mines, and brought to maturity by the automobile. From time to time, the curious would come to view the moonbow, especially when a big harvest moon came up of an autumn evening and couples in search of romance would snuggle under blankets with a flask and wait for the phenomenon to manifest itself.

Even with the moonbow, and the various parks and lakes that surround the town, and the proximity of the Cumberland Gap, through which Daniel Boone led his pioneers, Corbin has never been a powerful magnet for drawing tourists. For most of its history, the town has been more famous as a place people went through than one they went to.

When Americans first began to take to their automobiles for fun, many of them came through Corbin, along the Dixie Highway, from the Midwest and the North, en route to Florida. The Dixie Highway ran right through Corbin's main street and jammed it with tourists. People would stop and then later remember the little town. Corbin residents would themselves go to Florida and meet people from other parts of the country who would say, "Corbin? In Kentucky? That little town with all the traffic jams?"

In many areas, the Dixie Highway, officially designated U.S. Route 25, split into two routes when the association that marked it couldn't agree on which set of businesses to leave out. One place where this happens is just north of Corbin, where U.S. 25 West veered south toward Knoxville and U.S. 25 East bore off toward Asheville, North Carolina. At that fork took place one of the exemplary stories of road legend, tracing the change from Ma-and-Pa enterprise to modern corporate roadside business.

In 1929, a man from Indiana moved down to run a new gasoline station that Shell Oil was building at the fork. He painted his name and the word "Servistation" on the portico over his pumps. He joined the Rotary and the Kiwanis and the Masons and soon was so well established in town that everyone assumed he had always lived there.

He was a friend of some of the many good-roads men in town. He hated the railways, where he had his start in life as a stoker on the Seminole Limited of the Illinois Central Railroad. He left because of a tendency to get into fights with foremen along the lines.

He became known in Corbin as a charitable man, bringing ice cream to the Galilean boys orphanage one afternoon and feeding the boys until their bellies swelled. Occasionally he delivered babies in the poor mountain homes around Corbin. But he also had a short temper. The area around the station was known, like many another in the Appalachians and their foothills, as "Hell's Half-Acre," from its poverty and from the vicious fights that would regularly break out among bootleggers. Then the man would get angry and chase the bootleggers off his property with a rifle, shouting and swearing as he waved the gun around.

When the man decided to install a compressed air pump, he kindly had the installer run a line across the road to his competitor. The competitor, who spent most of his time strumming a mandolin and tending a pet pig, was less than ambitious and soon sold his station, which had a better position, commanding as it did the wide part of the curve, to the man who had given him the air.

After a while, the man began preparing snacks for his customers. Then he bought a sixteen-foot piece of linoleum "rug" on credit, laid it in a little storage room in the corner of his station, dragged in his kitchen table and some chairs, and called it a café. The man had learned to cook early in his life and knew how to turn out fine ham and eggs breakfasts, fried chicken and fried okra, lemon pie and pound cakes for dessert.

As time went on, the café became famous among tourists. The man spent hours driving through the adjoining country, learning as much as he could about road conditions—which were constantly changing as WPA-financed repair projects were finished—so as to advise his patrons, and scouting barns on which to paint advertisements. He chose barns because the area was full of hunters who shot at billboards. They would be more cautious if they knew that behind the sign there was a cow or mule whose owner might come after them.

He expanded the restaurant. He installed tablecloths, although they didn't always match, and flowers, glads from the sideyard. Sometimes the place took on the air of a Hopper painting: an eloping couple down from Nashville would sit alone, holding hands and watching the snow pile up outside. At other times, it was full and jovial, as when a well-lubricated carload of couples on the way back from a Tennessee-Kentucky football game would stop for dinner.

It was difficult, however, for travelers of the time to distinguish among the various Ma-and-Pa eateries. They quickly learned the foolishness of "following the truckers." They found out that down home advertising didn't always lead them to good food.

But in Corbin they found the real thing, a man who kept inquiring about how they liked the food, who would fly into a tantrum at the cooks and waitresses when anything went wrong, who installed a sign in Olde English script bearing the legend "Good will: The disposition of a pleased customer to return to the place where he has been well treated," who put on his menu phrases like "Country ham breakfast—$1.50. Not worth it—but mighty good."

He entertained the customers. He had a pet crow named Jim Crow that would pick pennies out of his cufflinks. He told stories about the local moonshiners and tall tales of his days working on the railroad and getting in fights with his bosses. When the audience was deemed appropriate, he poured over these stories, like gravy over his ham or chicken, a profanity whose color and inventiveness was remembered years later by his listeners. "He had a heart as big as a barrel," said a man who knew him then, "but, Lord, he would cuss a blue streak."

People remembered the owner and his stories and the good food and they would return year after year. When Duncan Hines, the critic of road food, who was from Bowling Green, Kentucky, included the café among the listings in his popular little book called *Adventures in Good Eating*, business boomed.

Eventually the man added a motel on to the restaurant, put shutters and flower planters on its windows, planted azaleas and dogwoods out front, deployed metal chairs in a variety of pastel colors on the little lawn, and Tudored up the whole place with fake half-timbers. Out front, the sign advertised tiled bathrooms, steam heat, and radios in every room.

Thus, by the time World War II broke out, the man, who was of course Harland Sanders, had become one of America's more prosperous roadside entrepreneurs. He had made some political contacts, and a few years before he had been named a "Kentucky Colonel" by Governor Rudy Laffoon—a wholly honorary but convenient designation that, of course, he was to exploit to the utmost.

After the war, he grew a little goatee and began to wear a string tie that seemed appropriate to his title. The image helped him when he chatted up the patrons. "Time to go out and do a little coloneling," he would say with a sign to his kitchen staff as he headed out into the dining room.

Fried chicken was not originally one of Sanders's best-selling items. He loved chicken and knew how to cook it well, but could not do it quickly enough or keep it fresh for long enough to make it an economical item on his menu. Good fried chicken takes at least half an hour, and, more realistically, forty-five minutes to cook.

In 1939 he had first encountered the pressure cooker. Experimenting with it, he found a way to cook chicken faster. He added it to his menu, scrupulously straining the oil after each batch.

He sought the perfect spice combination, mixing the spices on a clean concrete floor on the back porch of his home in Corbin, scooping a crater in a pile of flour and swirling in the spices.

The much-touted secret formula of "eleven herbs and spices" was not the critical element in the success of the chicken. It was speed in cooking, and constant filtering and changing of the oil to keep it from oxidizing. Packaging was also crucial: the regionalism, the Ma-and-Pa values, the quality and predictability, all summed up in the figurehead of the Colonel.

Then, in 1957, plans for the new Interstate system were announced, including the construction of I-75, bypassing Corbin. The town fought the move, appealing to state highway officials, but the officials knew about the famous traffic jams on the main street in Corbin, and the most they would do was move the Interstate a few miles closer to town.

Sanders had been bypassed before, when his Asheville motel and restaurant had been left high and dry by a new turnoff. Two years before the bypass, he claimed later, he had turned down $164,000 for his place. Now he auctioned it off for $75,000. Sanders was sixty-six. With the money he made from selling the restaurant and his $105-a-month Social Security check to live on, he set out in his Oldsmobile to salvage what he could of his capital—his reputation and his recipes. He began to try to sell franchises for his chicken recipe.

The Interstate system was to give a huge boost to franchising all over the country, strengthening American dependence on the car and thereby helping the older strips and in addition, with its intersections, practically mandating good locations for gas stations, motels, and restaurants.

Sanders gave away his first franchise to his friend Pete Harman, a Salt Lake City restaurateur with whom he attended religious meetings, and began to sell others. Sanders roamed the Midwest, cooking his chicken as a demonstration for the restaurant owners and, when he could, sneaking KFC place mats onto the man's tables. Some nights Sanders and his wife Claudia, one of his waitresses whom he had married in the late forties, would sleep in the Oldsmobile. Harman, who was a bit more conventional than Sanders, touted the chicken franchises at meetings of the National Restaurant Association.

This was not a slick corporate operation: Sanders continued to trade on his homeyness, his stories, his image. He put old friends of his, who had gone bankrupt, in the business. Later, he would come back to Corbin and give money to people in trouble.

The local franchisees made up their business as they went along. A Georgia outlet ran a promotion featuring Miss Georgia Poultry Princess. A Clarksville, Tennessee, franchisee painted his new Ford Mustang in the red and white stripes of the corporate image. A skydiver who landed in the parking lot was used to promote the opening of the KFC at Bambi's Motel in Griffin, Georgia.

By 1960 Kentucky Fried Chicken was available in 200 outlets; by 1963, it had 600 outlets and was the largest fast-food chain in the country, larger than McDonald's, larger than Shoney's or White Castle.

In 1964 Sanders sold out for $2 million to an ambitious young lawyer from a good political family named John Y. Brown and a wealthy investor named Jack Massey. Brown and Massey promptly began affecting string ties like the Colonel and touting the business with the energy of patent medicine salesmen.

They somehow managed to get the Colonel booked on Johnny Carson's *Tonight* show, where he appeared along with a plexiglass box containing $2 million in singles.

Over a period of just a few years, Kentucky Fried Chicken, like most of the fast-food, motel, and other franchises, changed from the most informal sort of family-and-friends-based organizations to slick, standardized outfits, with the number of steps between fryer and counter scientifically calculated to be the same in Paducah as in Portland, and with extensive national television advertising.

In 1971 the operation was purchased by Heublein, the liquor company, for $275 million. In 1982 Heublein was purchased by the huge R. J. Reynolds Industries, best known for its tobacco products but now a highly diversified megacompany. The Colonel became a lonesome figurehead in this corporate world. The middle managers had confused the man with the logo and forgotten that the Colonel was flesh and blood. At one point he had to threaten an embarrassing lawsuit to get the residuals due him for commercial appearances.

Everyone knows how Sanders became one of the most recognizable figures in the country, in the early seventies polling ahead of Richard Nixon and just behind Santa Claus in "familiarity factor." From Harland Sanders, Hoosier emigré, he had become the Colonel, and then, as he was known around the new Louisville corporate headquarters, "the old man," and still later, when he was little more than a logo, "the mug."

As he grew older and became ill with leukemia, his trips down the hall from his office in the big pseudo-antebellum mansion of a corporate headquarters to greet visitors at the Harland Sanders Museum grew less frequent. A life-sized plastic replica of the Colonel, like those in many KFC stores, was placed in the museum.

It was startlingly lifelike, with the string tie at just the same angle it assumed in every photograph of the Colonel, the glasses with the black top frames and arms exactly replicated, and the whiskers of the goatee pulled together in just the same way.

All his life, the Colonel said, he had "fought the curse of cussin'." "I knew the terrible curse of cussin' would probably keep me out of heaven when I died." He was quick to anger, and when he did colorful vocabulary invariably followed.

Then, in his seventy-ninth year, he got religion. He had always been religious, gone to church and tithed, but this was real, born-again, fundamentalist religion. He heard a Rev. Colman McDuff at the Evangel Tabernacle in Louisville, raised his hand, and went up to the rail. Not long afterward he was baptized again in the River Jordan. Jesus healed a polyp on his colon, the Colonel said. And He helped him stop cussin'. His conversion from downhome gas station operator to plastic saint, corporate front man, was complete.

Mansard roofs. The architectural sign of the franchise, a cheap and hospitable false roof, of plastic or wood shingle, with "environmentally sensitive" overtones. (POPEYES PHOTO COURTESY KIT MORE WOHL)

13: Franchised America

"It is ridiculous to call this an industry . . . this is rat eat rat, dog eat dog . . . I'm going to kill them before they kill me. You're talking about the American way of survival of the fittest."

—Ray Kroc of McDonald's

Like glaciers moving down from the north, the superhighways changed the ecology of the roadside, killing off the unadaptable individual enterprise and creating a climate for the growth of flexible but standardized systems. The superhighways and their bypasses worked quietly but universally to push highway enterprises toward standardization and franchising, both along the older roadsides and around the new interchanges.

In the new roadside environment, the old do-it-yourself enterprises—the giant ducks and pigs and coffee pots and pianos—were dinosaurs. What was required was a completely new mode of roadside reproduction, the sleek, swift, mammalian system of the franchise, in which the young were not simply hatched out into the cold, vicious world but suckled and nurtured to maturity by a parent organization.

The franchises were also better suited to the new environment in which advertising and real estate were the defining conditions of the economic climate. It was in 1957, just as the Interstate program swung into high gear, that the service sector surpassed the manufacturing sector as a proportion of the national economy; but the old, quasi-religious idea of personal service was replaced with the idea of service as efficiency.

The new roads had picked out sites for the franchises. They divided long-distance, tourist traffic from local traffic in many places. They fostered business travel, creating, in the form of the tired and dislocated salesman, a demand for nationally recognized, predictable bed and board, and a market in which to efficiently advertise it.

They encouraged tourism. The family vacation, in the era of the baby boom, was by force of economics an automobile vacation. People visited parts of the country they would never have seen before World War II—they wanted to see the country they had left to defend—and when they got there they wanted not only unfamiliar sights to see but familiar lodging and eating establishments to resort to—the comforts they had given up to protect. As a Ford ad of 1951 put it:

> Today the American Road has no end; the road that went nowhere now goes everywhere.... The wheels move on endlessly, always moving, always forward—and always lengthening the American Road. On that road the nation is steadily traveling beyond the troubles of this century, constantly heading toward finer tomorrows. The American Road is paved with hope.

The war moreover, had given the American melting pot a good shake. It had dispersed Appalachian farm boys to San Diego, Boston Italians to Albuquerque, Oregon lumbermen to Pensacola. It sent federal funds pouring into out of the way towns and cities and, after V-J day, lured many veterans to move to towns they had been introduced to by military service: Denver, Phoenix, Los Angeles.

But the superhighways also swept past thousands of small towns the old roads had built up, abandoning them the way the railroad had abandoned towns in the West. Businesses caught by the bypasses either moved and expanded or died.

The franchises that boomed in the late fifties and the sixties replicated the organization of the Interstates: a partnership of national standards and know-how with local capital and control. The franchise offered a combination of old and new forms of business analogous to the combination found in the food and ambiance of the restaurants: in franchising, traditional entrepreneurship blended with modern centralized management techniques.

The franchises were *kits*, assemble-it-yourself systems of products, finance, location, service, architecture, and advertising. The more flexible, the more neatly packaged and replicable the kit, the more successful the franchise. It was the ambition of every inventive restaurateur or motelier to create a chain. That is why the chili bowls, the teepee-shaped tourist cottages, the giant pig barbecues, the iceberg-shaped ice cream stands, the hot dog-shaped frankfurter stands, were all patented by their inventors. They expected the world to beat a path to their strange stands.

But the earliest chains, which expanded with their own capital, never extended to more than a couple of hundred outlets. The famous White Castle, born in the Midwest in 1921, which expanded without franchising, never boasted more than 175 units, almost all of them in the Midwest. With franchising, KFC reached that figure in just a few years. By 1985, it had more than 5,000 outlets.

With the leverage of individual capital, restaurants like KFC, McDonalds, and later such coldblooded creations as Wendy's (dreamed up in the early seventies by a professional fastfood executive) swept through the American strip like some new breed of plant, spread by windborne propagation.

Under most franchising agreements, the local franchisees, who pay an initial fee for their territories, agree to abide by certain national standards, buy supplies from the national company, pay a certain portion of their income to the national organization, and contribute a fixed portion—one or two percent of receipts—to a national advertising fund.

The product the franchise offered was not burgers or chicken but a whole package for selling those foods—image, methods, architecture, location. Kentucky Fried Chicken's product was not Harland Sanders's chicken with its "eleven famous herbs and spices" but speed and freshness combined with an aura of regionalism and personality. Ray Kroc's product at McDonald's was not the hamburger but the speed of service that kept the French fries fresh, and the predictable promise of the Golden Arches.

Many things had to be discarded in the transition from individual restaurant to chain. Colonel Sanders's gravy was one. Company officials said it was too hard to make. But for Sanders and millions of other farm boys who had grown up poor enough to appreciate gravy as a major constituent of their meals, gravy had a magic significance. As the years went by, Sanders criticized what the company had done to his gravy. "Ain't fit for my dogs," he would say. Franchising meant the end of gravy and all it stood for.

Roadside businesses faced the problems that any company, whether it makes computers, automobiles, or popcorn poppers, faces when it grows very rapidly: how to obtain capital for expansion without ceding significant control to outsiders, and how to maintain the entrepreneurial drive of its early days while enjoying the benefits of integrated administration.

Franchising seemed to offer a happy medium between that villain, "big business," and that frequent casualty, small business. It made a man his own boss without exposing him naked to the exigencies of the market. The franchisee was the driver, so to speak, but the franchiser made sure his course was well paved, well marked, and well adapted to the contours of the unknown country that lay ahead of him. The franchise packaged business opportunity as neatly as it packaged the hamburger. It provided an off-the-shelf business, a realizable ambition in kit form.

Eventually, the franchise firms adopted many management techniques most commonly associated with high technology companies. The common factor was dependence on labor—skilled in the latter case, unskilled in the former. The aim was to thwart unionization and defections of valuable employees by trying to foster a spirit of teamwork, family, value for individual "fulfillment," and so on. The gyms, saunas, and beer bashes of Silicon Valley were anticipated by McDonald's, where Transactional Analysis, with its "strokings" and "rap sessions" helped keep workers happy, and in its Think Tank, a windowless room with a huge waterbed where employees were encouraged to do their brainstorming.

Franchising had a long history. The franchise was a financial machine, a lever turning the power of a name or an idea on the fulcrum of local capital and entrepreneurship. Coca-Cola and other soft-drink companies had begun their rise to prominence through bottling franchises. But it was the automobile that spread this business form: to provide the sales and service the automobile required, most early auto dealerships were franchised, beginning in the 1890s.

Virtually everyone who came up with a roadside gimmick—the people who patented giant oranges or igloo ice cream stands—dreamed of turning his brainstorm into a chain. But simple expansion was difficult. Unfranchised enterprises got no further than had railroad restaurateur Fred Harvey, whose clean and efficient dining halls, filled with the wholesome "Harvey girls," were a remarkable prefiguration of the highway-style chain. Harvey House Restaurants, however, failed to expand much beyond the stations of the Santa Fe.

The first franchising of roadside enterprises was accomplished from the top down, by the major oil companies, even before the arrival of the superhighways. Around 1930 gasoline companies began to lease their buildings to individual proprietors for a percentage of the gross. The system provided greater flexibility in meeting local price wars and gave added incentive to the station operators.

At the same time, the petroleum companies established a pattern by standardizing and rationalizing the design of their stations. They did market research and found that ease of access, clean rest rooms, and brand recognition, in that order, were the factors most likely to draw a motorist off the highway. Then, in the spirit of consumer engineering, they hired industrial designers to build these appeals into the stations. Norman Bel Geddes designed a prototype for Socony-Vaccuum (later Mobil), with a large entrance lane, painted lines, and rounded corners, which drew the motorist into the station the way the streamlined structure of his Futurama exhibit drew the line of visitors into its belly. (Getting out of the station was harder.)

Walter Dorwin Teague created the memorable Type-C station for Texaco: a design grounded in the streamlined Moderne but so simple as to transcend it, and one that became irrevocably associated with the company. The Type-C building, like many stations, was covered in white porcelained panels—a suggestion on the facade of the clean rest rooms within, constantly inspected, the company advertised, by a "White Fleet" of traveling Texaco representatives.

Before franchising, gas stations helped bring a whole panoply of world architecture to the American roadside. There were Shell's shell-shaped stations, based on the company's pavilion at the San Diego Exposition of 1915. There were pagodas and log cabins, miniature Erectheums and Pantheons, prefab cottage stations and massive stone structures. In many areas there was also a demand that a gas station fit into the residential neighborhood—thus semi-colonial buildings in New England, stone ones in Westchester County, adobe in Santa Fe, and Spanish mission in San Diego.

But by the thirties, as drivers began to realize that there was very little difference among the gasolines sold by each national brand and as service became a more important component of the oil companies' business, the need was recognized for a station that established a strong corporate presence. It should be nationally standardized, efficient and up-to-date looking. Teague's Type-C design did this best.

By the late twenties, the gas station had become an American institution to rival the crossroads store. The phrase "Fill 'er up" became a cliché, along with the repeated image of the mechanic crawling out from underneath a chassis and wiping his greasy hands clean on a towel as he bent to address you through the window. And if you wanted to know what was going on in town, of course you stopped by the gas station.

There were some who had grander visions of the gas station. Frank Lloyd Wright saw it as an "advance agent of reintegration." "Each station," he wrote, "may well grow into a well-designed, convenient neighborhood distribution center naturally developing as meeting place, restaurant, rest room, or whatever else will be needed as the decentralization progresses and integration succeeds."

Wright, Richard Neutra, Rudolph Schindler, and even Mies van der Rohe tried their hands at designing good stations. Wright's first such project was intended as part of his Broadacre City. It featured cantilevered shapes and plenty of neon. In the fifties he did another design, this one for a Philips 66 station in Cloquet, Minnesota. Although intended as a prototype for a series of stations, only one structure was built. It featured a sixty-foot pylon shaped vaguely like the escape tower of a spacecraft soaring above a concrete-block and steel structure that blended Prairie-house style with inflections of Wright's style.

You might easily drive past this station without looking at it twice. To the odologist, it bears testimony to just how incidental the actual architecture of such facilities is to their program and function.

The gas station became a particular bugaboo of roadside beautification groups. Lady Bird Johnson called petroleum executives to the White House to complain to them about their stations. To John Kenneth Galbraith, who, ironically, had first popularized the notion of the "service sector," wrote that the service station was "the most repellent piece of architecture of the past two thousand years. There are far more of them than are needed," he went on. "Usually they are filthy. Their merchandise is hideously packaged and garishly displayed. They are uncontrollably addicted to great strings of ragged little flags. Protecting them is an ominous coalition of small businessmen and large. The stations should be excluded entirely from most streets and highways. Where allowed . . . there should be stern requirements as to architectural appearance and general reticence."

Some companies tried to beautify their stations by disguising them under woodsy, suburban, mansard-type roofs. Others sponsored a whole new generation of modern design, some of which were quite elegant. For Mobil, Eliot Noyes—IBM's favorite designer—created for a simple, International-Style brick station building, a series of internally-lit white plastic discs, placed on masts like umbrellas, above round, brushed-metal pump stations, their cylindrical shapes supposedly suggesting the barrel—a traditional symbol of the Mobil/Socony companies, although only a buff was likely to know that.

By the late seventies, the gas station as we had come to know it had practically vanished. 200,000 were wiped out by the gasoline shortages of the mid-seventies, and service had become self-service. The folksy, friendly, wise gas station owner was gone; now people pumped their own gas. The free windshield wash, even the "free air" of the pump that Harland Sanders and other get-ahead owners had made standard, vanished. You did your own windshield, and you might even be required to deposit a quarter in the air pump to fill a soft tire. The familiar grease pits and bays with hydraulic lifts were closed. It was no longer economical to "do service" at all.

The service station as a building type was completely rethought and redesigned for the new age of self-service and no maintenance. The clerk hid inside a bulletproof glass booth and remotely operated pumps with digital numbers. In Exxon's late-seventies design by Saul Bass/Herba Yager Associates, the familiar body of the station itself, with rest rooms, service bays, and office, vanished in favor of a simple roof over the pumps—often made into a dramatic, floating platform—and a booth for a single clerk who took the money and operated the pumps by remote control.

But the gas station had a successor, perhaps the "community center" Wright had envisioned, but in a form he would not have recognized. It was the convenience store, where gas was sold along with a simple array of groceries and such sundries as chewing tobacco, keychains, and caps.

With names like 7-11, U-Totem, Li'l General, Majic Market, and Quik Mart, the convenience store spread quickly through the American roadside. It replaced the Coke and Nabs machines of the old service station with a whole array of foodstuffs for the traveler, and enabled the local to drop in for bread and beer at the same time he filled his tank.

Of the major restaurant and motel chains, only Howard Johnson predated the war. He had made his mark first in New England—where the progress of road building ran ahead of the rest of the country—and then gradually in the mid-Atlantic and finally the southern states, along the Maine to Florida vacationland.

Johnson, however, was a model. His combination of colonial and Moderne summed up the aura the new traveler responded to, Mount Vernon's cupola rendered with the five-banded decoration of Deco. Johnson was the apotheosis of the Yankee peddler—he had begun as a traveling ice cream vendor—who valued cleanliness as next to godliness and the orange roof as at once a giant logo and a psychological incentive to appetite. From a single drugstore soda fountain in Woburn, Massachusetts, the man called "Buster" by his family franchised to thirty-five outlets by the mid-thirties and a hundred by the beginning of World War II. Howard Johnson's first restaurant benefited from the production in Quincy of Eugene O'Neill's *Strange Interlude*. Banned in Boston, the long play included its own interlude: a dinner break.

Johnson's traditional virtues had acquired an air of the up-to-date by careful promotion emphasizing the scientific production processes of his ice cream, which, by "a special formula"—all the restaurant chains would soon boast special formulas—contained nearly fifty percent more butterfat than his competitors'.

The science extended to the details of serving. Howard Johnson "Taylorized" the restaurant business, as the engineers were Taylorizing building and driving on the new superhighways. Johnson was the first to introduce a manual of standard operations—the "Bible," it was called—and the practice itself became standard in fast-food operations. The Bible stipulated that there be exactly 19–21 clams per portion; every hot dog was to be sliced exactly six times along its side; coffee was to be poured to just three-eighths of an inch from the top of the cup.

The packaging was as important as the food. The clams, for instance, arrived in a paper cup, fitted into a hole in a sort of metal palette that also held sauce and a napkin. The hot dog slipped into a stiff paper container decorated with the familiar logo and Simple Simon silhouette.

The whole treatment reflected moral fiber, cleanliness, efficiency, speed of service. All these values were combined in the person of Howard Johnson himself. He was a reassuring presence, a logo as comforting as the orange roof itself. His portrait was proudly hung in every restaurant.

To the marketing skills of the peddler, Johnson's operations added a form of New England rationalism, driven to an extreme, the rationalism of the clipper ship captain or of a captain of industry like Eli Whitney. Each operator was required to keep a daily log, like a ship captain, and the quality of franchises was monitored by the parent corporation's itinerant "shoppers." Lapses from standard could be reprimanded by the ruling council of franchisees, or "agents," back at headquarters, a group that *Fortune* magazine compared to "a New England town meeting."

The architecture of all the restaurants was created by a staff of twenty-seven at the central office, headed in the forties by a man named Joe Morgan, who made sure that the same elements—cupola and orange tile roof—were carried out in all sorts of styles, from Moderne glass block to nouveau Georgian.

Johnson built his largest outlet, a $600,000 luxury liner of a restaurant, with many bay windows and dormers, on Queens Boulevard, in 1939. Behind it were visible the Trylon and Perisphere of the World's Fair, on whose grounds he introduced his latest product: fried clams. These clams were emblematic of the way the franchised systems turned regional specialties into national obsessions. Visitors could repair from the dizzying visions of the future offered by the Futurama, their heads still swimming, to think out all the implications in an Olde New England setting.

While the fair was still running, Johnson obtained the restaurant concession for the new Pennsylvania Turnpike, the people's super-highway. The success of his operations there later established him on the Massachusetts, Connecticut, and New Jersey turnpikes as well. Ensconced in traditional-style Pennsylvania stone buildings, Howard Johnson's struck just the note of the Modern Olde the traveling public was interested in. Everything was just like colonial times—and right up to the minute; completely new—and just like Grandma made.

Howard Johnson's rationalized food operations were carried further by Ray Kroc and McDonald's. The story is well known: Kroc, a salesman for the Multimixer milkshake machine, became curious about a burger stand in San Bernardino, California, at the end of Route 66, that had so much business it required six of his machines. He spent hours in the parking lot watching the lines move in and out of the stand, calculating the daily and weekly gross. Then he spent weeks persuading the McDonald brothers—two Vermonters who had come to Hollywood and gotten into the film prop business before venturing into fast food—to sell him franchising rights.

Kroc rationalized the hamburger. His burgers were to be exactly 3.875 inches across, weigh 1.6 ounces, and reside on a 3¼-inch bun. Their fat content was to be no more or less than nineteen percent. They were to include no ground-up lungs, hearts, or cereals—this last dictum implying frightening things about the standard burger of the early fifties.

Kroc spent three months of experimentation to figure out why the McDonalds brothers' French fries were so good. It was the dry desert air of San Bernardino, he finally concluded, and he equipped his Midwestern stores with fans to produce the same effect, fans that ran constantly in warm store basements as blizzards whirled outside.

But it was a sign of the chain's dependence on the new highways that, by the mid-sixties, McDonald's single largest source of profit was not burgers or fries but real estate. Kroc had little interest in real estate, but his associate, Harry Sonneborn, pioneered a plan by which McDonald's owned most of its stores and leased them back to franchisees. The rental, along with the standard percentage of the gross, gave the company a return of around eleven percent. McDonald's found success because it was sensitive to the roadside environment: it chose locations carefully.

It was typical of the fate of many early roadside restaurateurs that, after making their deal with Kroc, they eventually lost their original stand. Kroc built a "real" McDonald's across the street from the original that preempted all their business.

A second generation of franchises were created differently: by cold-blooded entrepreneurs who saw an unexploited "market niche." David Thomas, a former manager of several Kentucky Fried Chicken outlets, built Wendy's from its opening in 1972 to a thousand-restaurant chain in less than five years. His gimmick was a larger, square hamburger to which customers added their own garnishes. Other developers moved to develop new gimmicks. Was there a movement toward leaner food? D'Lites quickly lined up investors for its "lite" chicken sandwiches. What about "down home" "country" cooking? Country singer Bill Anderson and others backed Po' Folks, with its biscuits and ham, and outlets shaped like share-croppers' shacks. Video games were hot? Nolen Bushnell, the inventor of Pong and founder of Atari, lured the kiddies by combining pizza with arcades full of video games and floor shows of cartoon character robots. He called the result Chuck E. Cheese's Pizza Time Theater.

Long before the motel there was the "tourist cabin," echoing Lincoln and the pioneer days as early auto travel echoed the migrations of the pioneers. Originally, motorists had stayed in pastures lent or rented by farmers. Then came civic campgrounds, set up by towns eager to draw tourist business, and finally private motor camps, with such amenities as water and restrooms. The evolution from pasture to motor court and motel, which took place in about fifteen years, from 1920 to 1935 or so, represented an evolution from a vision of meeting and mixing on the great public space of the road to one of family-oriented privacy and convenience. An economic segregation naturally accompanied this change. Auto tourism began with such romantic words as gypsying, hoboing, and bohemianism. It ended with a firm conviction that these represented sound values—as long as one did not have to rub elbows with any gypsies, hoboes, or bohemians.

The motel owner's biggest problem was scale: the bigger the motel, it turned out, the easier it was to survive the thin months. And that took a lot of capital, right away. Gradually adding units was difficult.

Another difficulty was the stigma that had burdened motel owners almost from the day in 1925 when James Vail of San Luis Obispo, California, came up with the word *Mo-tel*: a reputation for sin. The local "hot bed," "hourly rate," or "Mr. and Mrs. Jones" trade helped tide owners over tourist off-seasons, but it gradually hurt their general business. No less influential a figure than J. Edgar Hoover warned the public in the thirties that motels were hotbeds of crime.

Kemmons Wilson, the creator of Holiday Inn, had made himself several nice fortunes, in jukeboxes, construction, and other businesses before he got the idea of the chain. Supposedly, he was on a vacation trip with his family and was taken aback at the poor state of available roadside accommodations. With a tape measure, he personally laid out the ideal motel room.

Wilson's was not the first such effort. The Pierce Petroleum Company had built a chain of colonial-style luxury motor hotels beginning in the late twenties, but the Depression had shattered whatever chance they may have had. Operators of teepee-shaped motels—"Eat and Sleep in a Tee Pee," proclaimed the sign, which, as if the shape of the units themselves was not enough, was also shaped like a teepee—managed to sell several franchises.

Wilson's room was a definition of minimum comfortable living, a module like Le Corbusier's *Unités de Habitation*, which his later high-rise slabs resembled. It was also in the tradition of the American search for the basic living unit. Wilson paced off its proportions and requirements as carefully as Thoreau measured the dimensions and cost of his basic living space. As for architecture, who needed it? Holiday relied not on its building for image, but its sign. The best architecture was no architecture.

Of course the minimal standard has evolved: steam heat and tile baths, once luxuries that Harland Sanders could advertise on his sign, became standard. Television and then cable television replaced "radios in every room" and the even earlier "phones in every room."

Wilson, ironically, had a reputation for never getting a full night's sleep. He preferred to sleep in five-minute naps, sitting at his desk, riding in his limousine, or, he admitted, in church—he was a big churchgoer. These little modules of sleep were as crisp, as efficient, as representative of the modern businessman on the go, as Wilson's quintessential motel room.

The movement from family-run stores to a major corporate enterprise was reflected above all in the architecture of the chains. It was typical that Kentucky Fried Chicken's first director of store design was a man named Bill Bridges, whose "mother knew the Colonel's sister back in his home town of Henryville, Indiana."

Bill Bridges was working as an engineer in a cement plant when he was called in to bring some order to KFC's architecture. He looked desperately for some common image, finally painting red and white stripes on the diagonal because it was cheap and made the various types of buildings look like they had something in common.

At first, restaurants carried the chicken as just another item on their menus. As the chain began to expand, it was not enough just to put a sign out in front of any old diner or storefront restaurant that sold the chicken. For many of the new franchisees, the chicken—together with the trimmings, mashed potatoes, rolls, slaw—was now the main product. These were take-out stores. The new stores needed a "look."

Unlike McDonalds' or Shoney's stores, built from scratch, many KFC outlets were converted from old gas stations or dry cleaners. In 1961 a Kansas franchise was constructed in one week for $30,000 in a barn-style building. This was a record for the time, and earned the cover spot of *Drive-In Management*.

Local franchisees came up with their own designs. In the Carolinas there was a Mount Vernon knockoff—about thirty of them were built. How many contributions General Washington's casual design of his country seat has inspired along the road!—with that little cupola—so catchy, so classy, so cheap—and brick facing and picture windows in front.

Bridges seized on the bucket, the sort of oversized paper cup in which the chicken was packed—"Wow, a whole bucket full of chicken!"—as a sign logo. A Portland, Oregon, franchisee had first used it as an advertising symbol. Now, in 1961, Bridges created it in 3-D and motorized it as a sign. The first bucket was eight feet tall. But KFC was shipping these buckets out all over the country, sometimes five or six a week, so Bridges cut the size down to seven feet to fit common carriers like trucks and railcars.

The franchisees' magazine, *The Bucket*, advertised all sorts of decorative items for the local stores: the chandelier with four translucent buckets on it, the windvane made of silhouetted Colonel, complete with cane, and of course, the lifesize plastic Colonel, to be set in the corner. Then came the pagoda design, pioneered by a Georgia franchisee. The Colonel saw it and said, "Well, that looks like a damn circus tent."

From the early eighties, KFC stores were planned using computer-aided design. You feed in the site dimensions, estimate the traffic, add "the acceptable paraplegic arrangement," figure out how many Freshercookers you are going to need, locate the walk-in freezer, decide on a facing and roof style, pick a decor package—one is a sort of early American called "Heritage"—and the machine will spin the plans out onto the computer screen and then onto paper.

For one KFC national convention, Bridges and his staff built two entire model stores inside a hotel. The parts were delivered in four semitrailers, complete down to the kitchens and the landscaping in front. Accompanying each store was a videotape of the transformation of an older store to this "new image" store. One saw metal braces go up and prefab sections form into a mansard roof, aged brick facing cover cinderblock, a drive-in window grow like a bud from a hole in the side of the old store.

The new image store reduced the size of the pagoda to a sort of red-and-white-striped cupola or steeple, sitting atop the mansarded building. The stores looked for all the world like Southern Baptist country churches, rendered in up-to-date materials.

There was one thing that the architecture of all the roadside franchises, whether they were selling gasoline, burgers, or beds, had in common: lots of glass. The service station office, the McDonald's outlet, the Holiday Inn lobby, all had transparent walls.

Like the glass of the automobile itself, this glass mediated between private space and public. It almost seemed a form of reassurance to the motorist, looking at his car outside in the parking lot and at the stream of traffic on the road beyond, that by simply stopping he had not lost his mobility.

The universal architectural symbol of the franchise was the mansard roof. These mansards seemed to descend on American roadside buildings almost overnight in the mid-sixties. Some, like those installed by Kentucky Fried Chicken, went over older buildings converted to new outlets. Others were constructed as part of completely new buildings. Some were metal, like those on Ford dealerships, others were of wooden shingle. Some were of plastic molded in the texture of tile, others were Spanish tiles, like those on taco franchises. They were shiny or rough, orange or red, blue or yellow. They sat like little hats on cinderblock buildings or engulfed whole department stores, turning them into roadside mastabas.

The mansard roof was named for François Mansart, French Baroque architect of the seventeenth century, designer of the Château Maisons-Laffitte. The mansards, pioneered in the France of the Louis, were standardized along the boulevards of Paris, imported to New York during General Grant's administration, along with the whole affection for the Second Empire style, and revived in suburbia—French chateau-style houses—in the sixties.

The man who did most to make the mansard was Baron Haussmann, the Alsatian hired by Louis Napoleon, the creator of the Paris we know today. He was the Kissinger of city planning, vicar of a pasteboard emperor with a mustache and pointy little beard, who let nothing stand in the way of his widening the boulevards of the city.

Heights of buildings along the boulevards were carefully regulated, and the mansard was a way to tuck another story of space up under the eaves, for servants and poets. But the garrets of Paris became ancestors of the convenience stores of Amarillo.

The mansard also bore the ancestry of the Spanish-revival craze of the twenties—to be found not only in the Southwest but in twenties buildings in the Northeast—that had given us the first shopping centers, like Country Club Plaza—created by J. C. Nichols after a trip to Spain—in Kansas City or Highland Park in Dallas, those pseudo-adobe buildings with little tile roofs topping their thick walls. They even harked back to the little tin-roofed porches and overhangs of western general stores and saloons.

They were sheltering, not flat and open like the old drive-ins, because now you could eat inside the chain outlets. These shingled mansards, woodsy and rough and textured, were supposed to be country versions of a city innovation. They established a good will toward their suburban neighbors, the houses in the developments behind them, the customers who lived in those houses.

Howard Johnson's roofs, while not mansards, had become the symbol of the chain and showed the reassurance, the maternal sense of shelter, that the roof could symbolize. By the sixties, Howard Johnson's billboards showed nothing but the company's logo in huge letters, a location, and, at the bottom, an image of the roof, sliced off at the eaves.

In 1965 Lady Bird Johnson began her famous highway beautification drive, seeking to rid the country of the twin scourges of billboards and junkyards and urging each citizen to take personal responsibility for planting "a bush, a tree, a shrub" to make America a more beautiful place.

Shortly after the passage of the 1966 Highway Beautification Act, Lady Bird called representatives of the leading petroleum companies to the White House and spoke to them on the sore subject of the gas station and its ugliness. Her speech worked. Oil companies moved toward colonial or ranchstyle or Eliot Noyes-modern stations, and reformed old stations by applying mansards, much as KFC had applied them to the dry cleaners and gas stations it took over for outlets. Texaco covered a number of its classic, Walter Dorwin Teague Type-C stations with mansards glossed by vestigial dormers and iron railings. It did not even spare the little shelters over the pumps. Gulf mansarded its stations within an inch of their lives, installing orange roofs over not only the station and pump shelters but the very light stanchions and garbage cans. Before it was all through, Lady Bird had left a lasting mark on the American landscape and its highway culture.

McDonald's committed to the mansard in 1968, after having turned its arches from an architectural element into an icon. In some cases, the mansard grew so huge as to virtually swallow the building it roofed, like a too-big hat settling down on shorteared head. City outlets of the big chains created modified, flattened mansards to tack on the fronts of the row buildings they occupied. Mini-mansards grew up on garbage containers, over lights at gas stations, and topping off signs. Popeye's Fried Chicken went so far as to repeat the shape of its mansarded sign on its napkins, its boxes, and its cups.

Often, the mansard was open at the back, even though customers might drive around to park or order from the take-out window. The mansards hid the air conditioners and heaters and grease-laden vents from passing cars. Once you had them in the parking lot, however, it didn't seem to matter.

Shopping center by the freeway, Los Angeles, 1949. The highway shifted commerce from downtown to suburbia and then exurbia. (LOOMIS DEAN, © 1949 TIME INC.)

14: *The New Highway Landscape*

The supplanting of Main Street by the strip and of the wayside café or inn by the franchise were only the first great changes that road culture brought to the landscape. The freedom the freeway promised was not just freedom for traffic to flow smoothly but freedom from access. The changes turned the old, choked arteries the superhighways replaced into corridors for business rather than transportation, and new nodes of development sprang up at intersections and cloverleafs.

Roadside locations with access became premium properties, increasing many times in value as the Interstate and other new highway systems were constructed. In such high-priced locations, only the proven enterprise, usually the franchise, could survive. The building of the superhighways provided new patterns to the larger American economic and social landscape. The new highways worked in cooperation with the government's tax and loan policies, favoring single-family homes, and with the baby boom of the fifties, strengthening the suburbs the railroads had created and creating whole new ones.

If there was one thing above all that suburbanites demanded from the businesses that served them, it was parking. "Free parking" became a magic phrase along the suburban roadside. It was lack of parking that killed Main Street; Sinclair Lewis's characters complain that the spaces once sufficient for hitching horse and carriage were inadequate for the automobile. And it was in the most congested areas that new sorts of development surfaced to deal with the space the car demanded.

Thus the ubiquitous drive-in we associate with California has more to do with New Jersey, the state that has long led the nation in the numbers of miles of road per capita. The first drive-in theater, for instance, was opened on June 6, 1933, beside Admiral Wilson Boulevard in Camden, New Jersey, by one Richard Hollingshead, Jr. By 1958, their high point, there were more than 4,000 drive-in theaters; the energy crises of the seventies steadily reduced their numbers.

The term *drive-in*, as applied to restaurants with curb service and other such features, and to such enterprises as the legendary drive-in funeral parlors of Southern California, did not really become common until the late forties. Fast-food "drive-throughs," perhaps surprisingly, did not arrive until the mid-seventies, though they then flourished despite the fuel crises.

The central idea was that of being able to stay in the car, the next best thing to being able to stay at home. And in this, Southern California was the leader. When the Dodgers moved into their new stadium in Chavez Ravine, the thing that their deserted fans back in Brooklyn noticed first in the pictures was the huge acreage of parking. They had traveled to Ebbetts Field mostly by subway. Calling his Garden Grove church a "22 acre shopping center for God, part of the service industry," evangelist Robert Schuller listed the principles that animated his ministry—the same ones that drove the builders of drive-ins and shopping centers: "accessibility, service, visibility, possibility thinking and excess parking."

It was parking, too, that helped inspire the creation of the shopping center. Businesses along the strip began to locate close together, sharing parking, entrances from the road, signs, and customers. Put two strips together back to back, add a pedestrian mall in between, bring in a department store at one end, and you had the beginnings of the modern shopping center.

Two early manifestations of the shopping center appeared, naturally enough, in Los Angeles. The old Los Angeles Farmers' Market was really an open-air market with improvised roofs. Westwood Village of the late twenties was closer to the real thing.

The shopping center had other ancestors. The twenties saw the creation of the planned villages of Country Club Plaza in Kansas City, a vision of Seville in Middle America created by J. C. Nichols, who was to become a venerated figure in the shopping center world, and J. D. Prather's Highland Park in Dallas. In many cases, these developments marked a reaction by "upscale" retailers to the sprawl of the strip. Nichols made a study of shopping patterns around the country, tracing a pattern of development from the country store at crossroads to suburban groupings of stores to whole urban strips. "Throughout the land," he lamented in 1929, "the present-day congestion problem of our central business areas is being repeated in scores of outlying places throughout every large city." He planned Country Club Plaza as a response to these trends.

With the building of the new superhighways, the old main arteries became important chiefly as *locations*, as the business districts of suburbia. It was here that the shopping center, with its radical separation of pedestrian traffic and auto traffic and its scientific analysis of shopping—consumer engineering adapted to the new "discretionary income" of the fifties—found its natural home.

The prototypical modern shopping center that sprouted at cloverleafs, at the intersectional nodes of the new Interstates, and along the old boulevards and avenues—suburbia's main streets—was the creation of an Austrian emigré named Victor Gruen.

Gruen, an architect trained in the best traditions of the International Style, systematized the shopping center, helping to work out exact parking and delivery requirements and creating an architectural formula of square metal pipe, brick, and glass walls, derived from the basic International Style, that became the epitome of the modern center.

Gruen also introduced into the shopping center European-style amenities: cafés and plazas. He contended that the shopping center was a new building type, but with roots in the Greek stoa or Leonardo da Vinci's sketches of enclosed markets with service tunnels beneath pedestrian walks.

The difference between the new centers and their predecessors was evident in their names. The "plazas" and "villages" of the prewar centers gave way to "Northgates" and "South Hills"—names that were interchangeable with the names of the suburban developments they served.

Gruen advocated obtaining control of the zoning around a center—he did not discuss what political blandishments were necessary to achieve this—to provide the proper residential and commercial environment. He also wanted to control what he contemptuously called "pirating" stores, that would locate across the street from a mall and "ride its coattails" of parking and advertising.

Gruen's attitude toward zoning is suggested by an "invitation" to a public hearing his firm sent out before the construction of Southdale Center. At the hearing, the invitation said, "there will be a full presentation of Southdale and its importance to modern suburban living. The question of zoning will also be discussed." Southdale, in Edina, Minnesota, near Minneapolis, seemed to be Gruen's favorite among his many creations. At Southdale Gruen created the first covered mall by redesigning the galleria, beloved in Europe since the middle of the nineteenth century and briefly popular in America around the turn of the century. He took advantage of the problems of relating two magnet department stores and of the Minnesota climate to create the covered mall.

The first advertising for Southdale boasted that while the area had only 126 "ideal shopping days" a year, "in Southdale Center every day will be a perfect shopping day!

"You and Junior drive a few blocks to the Southdale Center. You park on the lower level, a short way from the entrance, walk in, and you're enjoying June in January!"

Southdale had pools with goldfish in them, a flashy symbol made of "S"s arranged in a star that appeared on everything from garbage containers to the uniforms of the center police, and a café. Two juxtaposed photographs in Gruen's book on centers, *Shopping Towns*

USA, are captioned "Dancing in the streets of Paris" and "Dancing at Southdale Center."

Through the work of Gruen and others, shopping center design became a science as exact as highway building. Using such formulas as "Reilly's Law of Retail Gravitation," and equations for figuring parking, careful traffic studies, and above all, consideration of the isochrones of the surrounding highway system, shopping center builders turned their business into a science. Their trade organizations—the International Shopping Center Association and the Urban Land Institute—published whole bibliographies of studies on such topics as the systematic deployment of store types—one for him beside one for her—channeling customer traffic by use of "magnet" department stores, security, eating patterns, video games, and the impact of the adolescents who, more and more, chose the mall as a place to "hang out."

Shopping center developers were fiends for statistics and traffic models. Their architecture was scientific, if often insipid, and they knew just who their customer was. "The compact plan and simple 50' wide Mall arrangement gives the shopper an easy grasp of where *she* [emphasis added] is going and what stores are in the project," boasted the architect of Mid-Island Plaza of Hicksville, Long Island.

Shopping center developers were unabashed about their intention of replacing Main Street. The stated aim of Mondawmin, a fifties center outside of Baltimore, one of whose creators was James Rouse, was "to recreate the atmosphere of Baltimore's Lexington Street [a major downtown shopping street] in a more pleasant, flexible, and convenient manner."

Many of the cities that boomed along with the superhighways, like Atlanta and Houston, were distinguished by their rings of shopping centers, arranged around the circumferential freeways—space stations in the orbital pattern. The Post Oak Road, Westheimer, and Galleria section of Houston, for instance, became the city's primary retail area, ranking well ahead of downtown in volume of business and in social cachet.

To compete, downtown merchants did their best to turn Main Street into a mall. The pattern of closing streets, paving them with brick or Belgian block, planting trees, and deploying benches is a familiar one. It was no wonder that Gruen's own ambitious plans of the sixties for recreating the centers of U.S. cities—a plan specified in greatest detail for Fort Worth—essentially called for turning them into super shopping centers. Pedestrian traffic would be restricted to a "podium" built on top of freeways and parking decks.

James Rouse, who became famous for Quincy Market in Boston and Harborside in Baltimore, carried this ambition to its logical end. In recreating downtown areas as shopping centers, he projected the center of the city as another concentration point in the whole urban matrix, one that, for shopping purposes, had the built-in advantage of a "theme"—historical and traditional—that fit in very well with the seventies shift of consumer attitudes toward the historical and natural. Even Main Street itself could be turned into a mall, even urban areas suburbanized.

The new highways created not just residential suburban development, with its supporting retail centers, but a whole new environment of office, laboratory, and manufacturing facilities—versions of the shopping center adapted for their own purposes. The enmity of highway planners to city life and the persistent American sentimentality toward the pastoral led to the recreation of the cities' functions in a new "ruburbia," whose bones were the new highways.

In these areas, the Interstates were viewed less as lines of transportation than as linear *locations*. The Interstates form the spine of most of the areas of speculative office, laboratory, and light manufacturing development that followed the movement of people to suburbia. Laid out like microchips on a circuit board, the buildings of these strips are most often associated with high technology; their best known examples are Silicon Valley, along Route 101 in California, and Route 128—actually I-95—around Boston. They are the chip strips, the new pattern of industrial and office development laid out by the highways. This development has brought its own characteristic architecture, a sleek, slick, glass-box treatment.

The patterns of Route 128 and Silicon Valley are replicated along U.S. 40 in central North Carolina, I-684 in Westchester County, I-276 near Philadelphia, on highways near Phoenix, Seattle, and Austin. Even along old stretches of U.S. 1, where it intersects with I-78 near Princeton, New Jersey, office park growth is creating a chip strip belt.

The chip strip form of development could adapt itself to any of the several geometries created by the Interstates. The first of these major forms was the spoke and wheel: Atlanta, for instance, created as a nexus of railway lines, was rebuilt in the sixties and seventies as a model Interstate city, with the routes meeting in a circumferential superhighway connecting the major shopping centers, industrial parks, airport, and other areas. Four or five freeways ran from the beltway to the city center, which was much smaller for the city's size than in a traditional city, and made up of hotels and convention facilities, several shopping centers, and a few office towers.

In Houston, a farflung city without zoning, built by energy prospecting and government spending on defense and space, not only the famous circular system of freeways but the long, open radii of access roads served as the armature for the city's sprawl.

U.S. Route 101, once the major north-south route on the West Coast, runs south out of San Francisco to San Jose as the Bayshore Freeway. The freeway begins from a standing start in downtown San Francisco, near the civic center, a monument to that city's prescient 1959 ban on new freeways, yielding its numerical designation to Van Ness Avenue. Through suburban San Francisco, it runs past Candlestick Park and the airport, through San Mateo and Redwood City and into Palo Alto, the northern edge of Silicon Valley.

Here begin the heavily developed stretches of the Valley: great ranges of low buildings, like so many computer chips on an integrated circuit board, occasionally interrupted open space, like the runways and ancient dirigible hangars of Moffett Field and, further south, the apricot and plum orchards that until a few years ago dominated the area's landscape. Before hitting San Jose, the road itself loses its median and interchanges to treacherous stoplights, famous for their casualties. Most rush hours, for much of its length, it is nearly impassable.

One can fairly trace the advance of electronics technology down 101, as the new firms spun off from old ones and sought their own locations. The process created a whole new organization of the landscape, one centered around the highway, and a type of architecture oriented to the freeway. These enterprises are famous for having started out in garages, and the architecture of the chip strips is of bigger garages—larger, nondescript, adaptable spaces, speculative buildings for speculative enterprises.

The office "chips" deployed along Route 101 are superficially varying but essentially identical "tilt-ups," their concrete walls poured flat in place, then raised and topped off by shingled mansard borders blocking the view of roofs filled with heating and air conditioning boxes. They are laid out like chips on a circuit board, "random access" buildings neatly linked by roads and surrounded by parking. With skilled employees at a premium in the Valley, the saying is that to change your job, you simply have to park a few spaces down, without leaving the industrial park. Linking the buildings together is 101, which might be described as the "power bus" for the valley.

This building type has a history. It began on the Stanford campus designed by Olmsted and Shepley, Rutan, and Coolidge—H. H. Richardson's old firm—in Palo Alto, where the Valley began.

With its courtyards and covered walkways, its low, Spanish-inflected roofs, the original Stanford spread out as lazily as the chip strips do, but on a pedestrian model. Additional buildings could easily be added to the plan. It was the planning equivalent of the parkways Olmsted had designed in the east.

The creation of Stanford Industrial Park in the fifties, which marked the birth of Silicon Valley, took this type of plan and made it even looser, providing plots like an oversized suburban housing tract. As companies became more successful, they "customized" their chips, building individual headquarters and labs that reflected a corporate image.

In 1953, Varian Associates commissioned Eric Mendelsohn, the great romantic expressionist of industrial architecture who had come to California in the thirties, to design its headquarters. The result was a departure from the curved concrete of Mendelsohn's romantic sketches. Boxy and rational looking, it restricted its romance to the facade and was to become the prototype of many International Style glass buildings that followed.

The best of the chip buildings along 101 were glass boxes, like the "black boxes" electronics types talk of, with an elegance of proportion and generous surrounding plantings. The ones built by speculators, on the other hand, tended to take on some shape distinctive enough to catch the eye from the freeway: rounded corners, or an oddly notched roof that served as a kind of visual logo for the development.

These buildings are designed for advertising themselves at superhighway speeds. Their architecture is slick but inexpensive. The dominant material is glass—not the clear glass of franchise restaurants or motels, but high-tech tinted glass, often partially reflective, so that it's almost impossible to look inside—a model of the confidentiality of "proprietary information."

People in the Valley talk about how the whole culture of the highway has affected life there. Automobile technology, for one thing, seems quaint in the world of electronics—a pattern of the past. They constantly use the beginnings of the automobile industry as a benchmark for the growth of their own industry. A high proportion of the technological whizzes value cars as antiques: they buy old ones and spend weekends fixing them up—sometimes in the same garages where they began their businesses. And they love high-priced new cars: the Ferrari dealership in Los Gatos is second in sales only to the one in Beverly Hills.

The individual values and choices of the freeway are an analogue of those of the entrepreneurial life. People in Silicon Valley believe that their area is full of risk takers because only the most mobile and ambitious people came west; the stick-in-the-muds were left home, sorted out by the process of the opening west. No wonder that an oft-repeated cautionary parable of the entrepreneur is the story of the computer developer who, on the afternoon the public sale of stock in his company made him a millionaire, died in the plunge of his Ferrari over an embankment near Los Gatos.

Route 128 in Boston, which eventually became I-95, laid out a pattern similar to U.S. 101 in semi-circular form. It was a whole highway planned as an industrial park, built as much as a pattern for location as a means of transportation. Advertising itself as "America's Technology Highway," Route 128 was created in the late forties by the stately Boston investment firm of Cabot, Cabot, and Forbes, and completed in 1951.

Regional planner Benton MacKaye, who called for "a townless highway and a highwayless town," had suggested a Boston beltway in the thirties. His plan was for a wide parkway, its two sets of lanes separated by a park, with more park land to either side—a greenbelt in the best traditions of progressive city planning.

Cabot, Cabot, and Forbes did it somewhat differently. In place of the recreational park, they created an industrial park; and instead of a parkway they persuaded the state to build a straightforward, no-nonsense expressway, since widened and rebuilt. The result was an odd juxtaposition of old New England countryside—village greens and stone walls—with the high-tech belt. Route 128 runs a few feet from the green in Concord where the Revolutionary War began. Firms along 128 have in some cases rehabilitated and moved into old, abandoned textile mills—some of the earliest American industrial developments—as the Wang Computer company did in Lowell, Massachusetts.

Some 900 high technology companies had located along 128 by the early eighties, and traffic had begun to thicken and slow. Route 128, in the term favored by real estate men, was becoming "saturated." The demands of the development had begun to exceed the switching capacity of the "bus." Development began to shift outward, to I-495, a second beltway twelve miles further out of Boston. The signs changed from "America's Technology Highway" to "America's Technology Region."

Freda Farms Ice Cream stand, Berlin Turnpike, Connecticut, 1940. Even before the big duck, this famous construction of superscale ice cream cartons— part Palladio, part Pop—was not only beloved of New England tourists but praised by architecture critic H.-R. Hitchcock. (LIBRARY OF CONGRESS)

15: *Why a Duck?:*
The Redemption of
Roadside Architecture

Ever since the early days of autocamping, intellectuals had condemned the roadside strip as the dreariest product of American
culture, all kitsch and chaos, "banality" and "sprawl." But by the
end of the sixties, as the superhighways turned strips into back-
waters, something odd happened: the strip became camp.

For architectural theorists, strip architecture by the mid-seventies
was a fashionable new "vernacular," replacing the earlier vernac-
ulars of grain elevators, Bucks County barns, and Shaker furniture.
The establishment of the superhighway network had made the old
strips food for nostalgia. And theories about roadside architecture
were to have a dramatic influence on "mainstream" architecture.
The favorite monument of the roadside as camp was located not
along Route 66, not in California, and not anywhere along the great
western roadside. It was not Mammy's Cupboard in Natchez or
the Brown Derby. It was a concrete duck, designed by one Martin
Maurer and built by two brothers named Collins, along Route 24,
near Flanders and Riverhead, New York.

Why did the Martin Maurer duck building become the most
famous, most talked about roadside building in America? Why not
the giant barbecue pigs of Texas, the drive-through donuts of California, the orange-shaped orange juice stands of Florida, or the
shell-shaped gas stations of North Carolina?

Why did the duck, with its eyes made of Model-T taillights and
a door in its breast, the duck that had been featured as a do-it-
yourselfer's triumph in the November 1932 issue of *Popular Mechanics*, winner of the Atlas Cement Company's "Most Spectacular
Piece of Cement Work for the Year 1931" award, become the focus

of a whole tempestuous theoretical exchange in the teapot of *Architectural Forum*?

The reason is simple: the duck was conveniently located to be taken up by the New York art and architecture world. The Big Duck was located on the way to the fashionable and originally artsy resorts of Long Island—Montauk and the Hamptons, Sag Harbor and Amagansett.

The duck stood on a stretch of road not far from where Jackson Pollock wrapped his car around a tree—and one sometimes traveled on weekends by a New York professor and architectural critic named Peter Blake.

Blake was the author of *The Master Builders*, a paean to the classic modernists, Mies, Corbu, and Wright, and of *God's Own Junkyard*, a volume expressing dismay at the visual state of the American landscape. He especially loved that part of Long Island where the duck was to be found, and he hated seeing the great expanses of its potato fields, the sudden prospects of the water that appear between the foliage, messed up with tacky commercial roadside architecture.

In that book, which became popular in the days of Mrs. Johnson's beautification program, Blake included a picture of the duck as a terrible example of roadside schlock. But influential architect and theorist Robert Venturi claimed that the duck picture and many of the other pictures in *God's Own Junkyard*, didn't look so bad after all: "The pictures in this book that are supposed to be bad are often good. The seemingly chaotic juxtapositions of honky-tonk elements express an intriguing kind of vitality and validity, and they produce an unexpected approach to unity as well."

Venturi was picking up on H.-R. Hitchcock's argument of the thirties that there was the glimmering of some new, future, "high" architecture in the architecture of the roadside. Hitchcock, in a famous aside near the end of a book on H. H. Richardson, had called the combination of symbolism and functionalism found in roadside architecture an "encouraging sign" for architecture. Without deigning to name it out loud, Hitchcock described one of the wonders of the old Berlin Turnpike in Connecticut, the Freda Farms ice-cream stand, with its oversized versions of ice cream boxes arranged like the masses of a Palladian villa. He added, con-descendingly, that "obviously, ice-cream boxes with silvered sundae dishes . . . filled with canvas strawberry ice cream and provided with ribbon windows for outside service, are incidental to our culture. But the very fact that those who condemn them as breaches of 'taste' prefer anachronistic Colonial or Tudor filling stations is proof of their symptomatic actuality."

Venturi, naturally, saw that glimmering of "symptomatic actu-ality"—whatever Hitchcock meant by the term—as a reflection of his own ideas about architecture, which were to become funda-mental principles of "postmodernism." The great thing about the architecture of the roadside was its combination of the functional so cherished by modern architects and the symbolic so long in disfavor among them.

Venturi saw the elements that he favored in high architecture, "contradiction and complexity," in the architecture of the strip: "contrapuntal relationships, equal combinations, inflected frag-ments, and acknowledged dualities." It was through Venturi's com-ments that the duck became one of the most famous, most argued about buildings of modern architecture.

The duck's primary attraction to Venturi, and others, of course, was that it seemed daring to refer seriously to such a pedestrian piece of architecture. Modern architecture was never possessed of much of a sense of humor, and the duck became an element not in a debate about roadside architecture, but about conflicting theories of all architecture.

For Venturi the duck was a weapon to beat over the head of modern architecture. He contrasted the duck with another building type, "the decorated shed," which has its archetype in the Western false-front. The buildings of modern architecture are ducks, Ven-turi said, not pure, ideal, functionalist shapes. When modern build-ings take on the swooping forms of flight in airport terminals they are representing the business of the building the same way the duck does. And even when they display their beams and girders as "pure form" and "pure functionalism" they are representing the business of the building: to be taken seriously as modern architecture.

The open girders and rough concrete of these buildings were

simply ornament posing as expression of structure. In many cases, Venturi showed, these elements didn't really hold anything up. There were no purely functional buildings, any more than there are pure, Platonic things or Kantian things in themselves.

Venturi did not argue in favor of reproducing the sprawl and chaos of the commercial strip, despite its charm. It was not all right, but *almost* all right. It still required an ironic architect to "redeem" it. "Is not Main Street almost all right?" ran the critical paragraph of his argument in the 1966 *Complexity and Contradiction in Architecture*. "Indeed, is not the commercial strip of a Route 66 almost all right? As I have said, our question is: what slight twist of context will make them all right?"

To call the architecture of the strip "vernacular" was to place it in a protective category, providing ironic, prophylactic distance. To the bourgeois gentlemen who ran all the Sugar Shacks and Dairy Freezes, who bought into Bojangles and Steak & Ale franchises, the term would have come as a surprise: to think that for all these years they had been speaking prose! And Venturi was almost as condescending toward roadside architecture as Hitchcock. It was like the masks in Picasso's Cubism: useful as part of the collage, but made by savages.

Twisting the context, Venturi hoped, would "rehabilitate" the architecture of the strip—"the commercial vernacular." Venturi wanted to employ this vernacular as a vocabulary for architecture that aspired to art, the way the modernists rehabilitated industrial architecture—"the industrial vernacular" of grain silos and Kahn factories.

Venturi never managed to do what he theorized. He designed a BASCO warehouse with letters as big as a house outside, and a building for Best Products decorated with supergraphics of giant flowers, but these were simply slick examples of commercial architecture. He never really managed to "change the context."

It was left to another admirer of the duck to carry out the program Venturi had set forth. James Wines, a man perhaps best described as a conceptual artist operating as an architect, replied to Venturi's articles and books on the duck with a 1972 piece in *Architectural*

Forum in which he argued that "form follows fantasy not function." For Wines the duck was an example of "individualistic architecture" that, by contrast with rigid, dry modernism and erudite, referential postmodernism alike, indulged in fantasy.

Wines was becoming famous for the roadside buildings he designed for a discount retailer called Best Products. The standard Best Products outlet had been a simple box, a sort of giant vending machine. You looked at display models of the products for sale, and the clerks brought them out from a storeroom at the back.

The program for these buildings never changed, but the facade could. So Wines broke up the facade. For a Houston outlet, Wines designed a building with a cracked facade and bricks tumbling down. For one in California he put the whole corner of the building on tracks so that a giant notch moved out each morning when the store opened, then moved back in at night. In another store he put trees in the middle, like a giant terrarium.

Wines set himself in relation to the modernists much as Jasper Johns set himself in relation to the abstract expressionist painters. The duck for him is not art but "an artifact." Venturi had talked about changing the context of roadside architecture, and Wines set out to do just that: to create art by changing the context of artifacts, playing off the expectations they bring with them.

In 1984, Wines designed an "exploded" version of the classic mansarded McDonald's for a shopping center in Berwyn, Illinois. He set all the parts a few feet from each other, walls slightly apart, roof lifted, as in an expansion diagram, to show the design for what it was: a kit.

Wines's office, which is packed with volumes by phenomenologists and structuralists, by Bruno Zevi and Ferdinand de Saussure, is located in what he will proudly tell you is the only building in New York designed by Louis Sullivan. He resents having his architecture "lumped with the duck."

"I use roadside architecture as a springboard. The duck is an artifact, not art. I make art out of artifacts.

"It's like the difference," he says, "between Neil Simon and Beckett.

"Roadside architecture is a perfect place for setting up semaphores. You've got a tremendous amount of reflex identification. Because of the high level of expectation you can trigger people's thought processes. People driving along a strip are very malleable."

Wines says that he really does not particularly like roadside architecture. He sees it only as providing a set of clichés, a field of expectations, that can be manipulated to produce art: artifacts suitable for turning into art. He is a conceptual artist who works the medium of architecture—a process he likes to call, with a nod to the European structuralists who adore his work, and who specialize in breaking down, "deconstructing," the "sign systems" of art— "dearchitecture."

Sometimes Wines can sound as romantic about art as any art-for-art's-sake proponent. "Art educates on its own," he will say. "It pulls people in. The guy with the shopping center next to our McDonald's has now got real problems. He's going to lose customers."

In fact, however, the most interesting thing about his McDonald's was just how indistinguishable it was from any other. You noticed the cracks in his Best Product facades, but the raised roof of the McDonald's, the slightly offset walls, simply seemed another of the thousands of possible variations the basic McDonald's mansard design allowed. It was a testimony to the power of the kit.

And that power, so essential to the second generation of roadside architecture, the architecture of standardization and superhighways, was what the new roadside campers missed. They appreciated the do-it-yourself folk art of the Brown Derby, which Susan Sontag included as an example in her famous essay, "Notes on Camp," but not the corporate folk art of the McDonald's kit.

While architectural high culture had hoped to absorb and "subsume" road culture, it was more nearly true that the opposite happened. High architecture, instead, began to mimic the accents of roadside architecture. The buildings in which Venturi attempted to "change the context" turned out to be simply more elegantly proportioned roadside buildings, and Wines' manipulated roadside artifacts, for all their good humor, remained artifacts.

Roadside culture resisted being carried home to academia, condescended to with quotation marks, replanted in some theoretical terrarium. It showed itself to be not a dialect, but a full-fledged language, different from the language of refined architecture, speaking in the shrill but energetic tones of advertising, at the raised volume not of the pedestrian but of the driver. Its "actuality," as H.-R. Hitchcock had called it, was neither "incidental" nor "symptomatic," but almost universal.

Philip Johnson, the architect who had helped create the 1932 International Style exhibition at the Museum of Modern Art, who had worked with Mies in the design of the Seagram Building, became a convert to postmodernism. He also became a camp follower of the roadside connoisseurs, praising and helping subsidize John Margolies's photographic surveys of old diners, drive-in theaters, and motels.

When Johnson was called on by AT&T to design a headquarters tower in Manhattan, he employed one of the most basic of roadside icons: the pedimented top. Critics talked about the classical references of the form and its resemblance to Chippendale furniture, but rarely noted its relationship to the ubiquitous broken pediment of strip signs.

Nor did they note that the best place—in fact almost the only place—where it was possible to see the top at all was not the crowded streets of Manhattan below the building but across the river, on the Long Island or Brooklyn-Queens expressways. It was too bad the building could not have been in Houston, where it would have been obvious that, like all the skyscrapers there, it was really sculpture for the freeway—superscale roadside architecture.

These were not the thoughts of every viewer of the AT&T tower, but few of the three million tourists who annually visit the Gateway Arch in St. Louis, Eero Saarinen's masterpiece of expressionist modernism, could fail to note one thing about it. This monument to Jefferson's Louisiana Purchase and to westward expansion, this triumphal arch honoring mobility, looked like nothing so much as a giant McDonald's arch that had lost its mate and faded from gold to silver. Lest anyone miss the analogy, there is a McDonald's franchise to jog his imagination just next door, established in a remodeled river steamboat.

The lure and promise of the open road.
Hitchhikers along a southern road. Photo by
Walker Evans. (LIBRARY OF CONGRESS)

FOUR

Gone to Look for
America

The Cadillac Ranch, monument to the mythology of the open road, along Route 66 near Amarillo, Texas. Created by the Antfarm art collective: Doug Michels, Chip Lord and Hudson Marquez, c. 1974. (COURTESY DOUG MICHELS AND STANLEY MARSH 3)

16: Sentimental Journeys: The Myths of Route 66

"If at the end of a year of unremitting labor [the American] finds he has a few day's vacation, his eager curiosity whirls him over the vast extent of the United States, and he will travel fifteen hundred miles in a few days to shake off his happiness. Death at length overtakes him, but it is before he is weary of his bootless chase of that complete felicity which forever escapes him."

—Alexis de Tocqueville,
"Why the Americans are So Restless in the Midst of Their Prosperity," Democracy in America

When roads were assigned their official U.S. route numbers in the mid-twenties, *The New York Times* editorialized that they had lost their romance. "The traveler may shed tears as he drives the Lincoln Highway or dream dreams as he speeds over the Jefferson Highway, but how can he get a 'kick' out of 46 or 55 or 33 or 21!"

The editoralist, of course, had missed it by a mile: travelers were to get their kicks on 66 for the next half-century. Sixty-six became a romantic designation. It was the highway that stood as the central route of a series of legends joined in a common myth: that you could find the Real America on the road, that to drive cross-country was to get a cross section of the country.

Sixty-six, of course, has always stood for a whole class of roads. The Okies didn't all take 66, they also took U.S. 54 and U.S. 80. Tourists took a number of roads to California. But 66, perhaps due simply to its mellifluous (and quite accidental) numbering, is the road that named the myth.

The highway that became known as Route 66 roughly followed the course of the old Osage Indian Trail. As a major emigrant route to California, it was known as "the Wire Road" for the telegraph lines along its course.

Sixty-six flowed conveniently from Chicago through heartland America, past corn and wheat fields, beside towering grain eleva-tors, through Missouri and Oklahoma and across the Texas pan-handle, into the real west of Indians and mesas, cattle skulls and wrecked Model T's beside the road, past Acoma Pueblo, the oldest continuously inhabited spot in the nation, into Arizona.

Before entering the Promised Land of California, 66 wandered with Biblical fitness through the wilderness of the Mojave, all sand and creosote bushes burning in the heat. The road here was nothing but planks until the early thirties. The Okies usually took it at night. Finally 66 reached the edges of greater Los Angeles in San Bernardino, where McDonald's was to get its start, and kept its number through the city as Santa Monica Boulevard.

The history of the myth of Route 66 is the history of our changing notions of the road as the way to the heart of America. At first, 66 was a road for the pioneer auto tourists in search of American wonders. Later, for the Okies, it was a route for playing out once more the hope of prosperity down the road, of the realization of the American economic dream, thwarted on their home territory but alive, surely, further west, in California. For the students of

America, journalists, novelists, and photographers, who followed the migration, it was a way to see what was really happening in the America of the thirties. By the fifties, it had become a road on which to see the dark and curious sides of America with an alienated eye, or again, a road to California—whose promise was still prosperity, but now compounded of blonds, surfing, and beach houses— in a Corvette. "Get your kicks on 66," the phrase from Bobby Troup's song had it; the width of that sentiment's appeal was indicated by the fact that musicians as diverse as Nelson Riddle and the Rolling Stones recorded the song. In the seventies 66 was a locus for collectors of the camp imagery of the roadside past: the teepee motels, diners, Indian Cities, and giant dinosaurs. Sixty-six stood for the whole complex of backroads culture.

It was Steinbeck, of course, who established himself as the Bard of 66 with *The Grapes of Wrath* and established 66 as "America's Main Street." But even the Okies that he and the journalists of the trek chronicled were nostalgic: they were reenacting a western pilgrimage. "Covered Wagon, 1939 style" read the caption beneath a photograph of a loaded automobile in Dorothea Lange's *American Exodus*. But the point of Steinbeck's book was that there was no longer, if there ever had been, an El Dorado, a golden orange grove, at the end of the road, only Bakersfield. The end of 66 was more driving—on the freeways of Los Angeles, where migration was stylized into circulation.

Architects and academics with a campy curiosity about and fascination with road culture saw it summed up in the myth of 66. By 1968, Robert Venturi was referring to Route 66 as the archetype of the open highway with occasional commercial strip, as we had once come to refer to Main Street as the quintessential small-town thoroughfare.

Writers and photographers attached their affections early on to Route 66 and the project of the cross-country, cross-section trip it stood for. Steinbeck and Dorothea Lange followed the Okies. The Beat Generation followed its restlessness back and forth, bouncing from coast to coast. Their souls, as Jack Kerouac wrote of Dean

Moriarty, his fictionalized Neal Cassady, were wrapped up with "a fast car, a coast to reach, and a woman at the end of the road." Pure existential restlessness was their thing.

Their successors, the Merry Pranksters of Ken Kesey's Magic Bus, equipped with fancy stereo equipment, with lysergic acid to replace the Benzedrine and booze of the Beats, recapitulated the journeys Kerouac and Cassady took in their old Hudson. At least one old Beat even came along: Kesey navigated but Neal Cassady literally drove. "Further" said the destination plate above the bus windshield.

It was clear very early that the automobile would change the way Americans looked at their landscape. Henry James, having returned to America from his long English exile in 1904, traveled through New England by automobile and wrote of "the great loops thrown out by the lasso of observation from the wonder-working motor-car"—perhaps the only time the Master cast himself, however ironically and distantly, as cowboy.

He compared the closeness and variety of the views offered by the automobile with the restricted views from the window of the railroad car. But the motorcar showed him almost too much of America. After his tour of New England, where his Anglophilia was less discomfited, and, more briefly, a trip southward to Florida, James simply gave up: he couldn't quite deal with the whole vast and changed expanse of the country. He was never able to complete the planned second volume of *The American Scene*.

The project of seeing the country was part of a restlessness as old as the country itself, but the automobile had changed its terms. The project remained, as it had for the builders of the Lincoln Highway, bathed in a nostalgia that shifted the objects of its affection with the course of time. It was tourism in the rear-view mirror.

Tourism was a search for the old values. It was more than simple recreation, it was a way to investigate as well as indulge the national restlessness. It was a way to see America, the real country, populated by "real Americans." "See America first" was the motto of the auto tourist. See America, that is, not before you see Europe, Asia, or Africa, but before your neighbors can see America. And see America, the proponents of the highway seemed to mean, as if you were the first to see it: regain the sense of discovery of the first explorers and pioneers.

Tourism on the roads became a scale model of the process of exploring the continent. No wonder that cars began to bear the names of explorers: De Soto, Cadillac, Hudson. The modern auto tourist, whether poet or vacationing accountant, saw himself in the tradition of the explorers, wishing to encounter, as they did, what F. Scott Fitzgerald called "the fresh green breast of the New World."

For the middle and upper classes, the car and the new roads provided a new form of recreation: autocamping. With a tent or a fly attached to the back of the roadster, millions of Americans headed for the countryside. Autocamping became to the early twenties what the fifty-mile hike was to the early sixties or the mass marathon to the late seventies. Autocamping offered the driver an opportunity to feel himself king of the road. As Frank Brimmer, editor of the autocamping magazine *Outing*, wrote:

> the car or trailer is M'Lord autocamper's castle . . . and the whole wide world is his manor! The autocamper, properly equipped, is a petty feudal monarch in a horizon that is all his own. On his rubber-shod castle grounds, with all the *lares et penates* of his home hearth carried in with him, he may set up roadside housekeeping anywhere on God's green footstool, free and independent of the whole wide world.

President Harding went on a famous autocamping expedition with Henry Ford, Thomas Edison, John Burroughs and other notables. Ford chopped the wood, and the party, with servants in the background, posed happily for the camera beneath a tent, arranged around a food-laden Lazy Susan.

Auto tourism was emphasized as a way to strengthen the family. Henry Ford talked of his Model T as a means for any American of reasonable means to enjoy "the Sunday drive," and take his family into "God's great open spaces." Autocamping gave the tourist almost complete freedom. As one autocamping booster noted, "you are limited only by the quality of the roads."

The improved quality of roads gave the auto tourist an easier time getting there, but the dubious nature of roadside food and lodging and of travel conditions continued to make travel an adventure. Especially for the tourist, the experience of the road was a constant conflict between comfort and adventure.

In the years before the superhighway, the road had been a dark and lonely place for him as well, even if the dangers he faced grew less intimidating over the years. Mud and dust, for instance, soon went the way of Barney Oldfield; but night driving was a novelty until the thirties, and risky for years afterward. Even as pavement spread across the country, the middle-class tourist found a new class of worries: the danger of strange places, strange people, and strange food, the danger of the unknown for the vacationer, venturing further afield, of wrong turns, crooked mechanics, and speed traps, of backwoods southern justices of the peace holding drivers for ransom in desolate small-town jails.

Roadside commerce constantly ministered to this fear. It introduced standardized national products to soften the strangeness of an unknown locale. It was reassuring to find Coca-Cola in the vending machine at the motel, Simmons Beautyrest Mattresses on the beds, and Congoleum on the floor. Roadside business also worked to turn regional variants into attractive, cartoon images. If the railroad created nations in the nineteenth century, in the twentieth the highway created a new notion of regions. Local areas were forced to present themselves to outsiders along the side of the road. In this self-presentation, the region became an abbreviated set of local clichés: the Dutch mill, the southern Doric porch, the southwestern mock-Alamo motel. And finally, of course, roadside culture eliminated regional variants almost entirely in the abstract consumer world of national products and chains. The need for comfort finally won its long battle with the desire for adventure.

It was the memory of old dangers, perhaps, that sharpened subsequent memories of an ideal roadside past, of the magic diner or café with the sun streaming through the windows and the breeze through a screen door with a Rainbo bread sign, where the taste of homemade apple pie blended with the constant rush of the passing traffic.

There were memories of the gas station with the huge red Coca-Cola locker, with walls as thick as a bank safe's and, outside, a bear, slunk in the corner of his cage as an attraction for the kids. Or the lonely motel room, with its universal scent compounded of carpet cleaner and disinfectant, its anonymous furniture and fake oil paintings. Such impressions became almost racial memories for the generations born into the auto age.

The boom in auto tourism saw the creation of the specially adapted roadside tourist attraction, which had no reason for existence except to attract tourists. The principles of the highway tourist attraction, established in the twenties, remained consistent even later, with the arrival of the Interstates. The most critical part of a spot's attractive power was the radiating pattern billboards that drew tourists from miles away.

Near Chattanooga, there was the Confederama, little more than a chairlift and a diorama of the battle of Lookout Mountain—"Completely Air-Conditioned," "No Walking," "Since 1957." There was Rock City, which used innumerable barn roofs as its billboards, and Indian City in Arizona, a souvenir shop in a geodesic "hogan." In Florida there were half a dozen Fountains of Youth and as many Gator Farms. Everywhere, from New York State to Washington State, there were the Western towns, with mock gunfights and customized "Wanted Posters" for sale.

On the road, its boosters argued, the motorist could find real Americans. The gas station attendant, the diner cook, the mechanic, all became wise voices in this search, roadside oracles. In 1939 FDR himself advised a young man to learn about the country by driving cross-country:

> Take a second-hand car, put on a flannel shirt, drive out to the Coast by the northern route and come back by the southern route....

> Don't talk to your banking friends or your chamber of commerce
> friends, but specialize on the gasoline station men, the small restau-
> rant keeper, and farmers you meet by the wayside, and your fellow
> automobile travellers.

The gas station operator figures again and again as the sage of
the roadside. Henry Miller, returning to the United States to produce *The Air-Conditioned Nightmare*, found a hero in the mechanic
who kept his aging heap running. Sinclair Lewis once advised a
young writer to seek material for his novels by talking to the local
gas station owner.

For Steinbeck, the heroic figures were the people who shared
food, water, shelter, and gasoline with those less fortunate, "The
good common people," Ma Joad says. "If you're in trouble or hurt
or need—go to poor people. They're the only one's that'll help—
the only ones."

The good people are also waitresses with golden hearts and truckers with charity beneath their rough exteriors. Steinbeck's waitress,
Mae, sells nickel-a-piece candy to the Okie children for half a cent.

Such figures remained part of the mythology of the open road.
A 1985 *Time* story about an Interstate truck stop told a similar
story. Truckers tease an exhausted waitress, who drops her tray
and bursts into tears.

"'I'm so tired, my old man . . . he left me . . . The judge says he's
going to take my kid away if I can't take care of him, so I stay up
all day and just sleep when he takes a nap and the boss yells at me
and . . . and . . . the truckers all talk dirty . . . I'm so tired.'

"She puts her head down on her arms and sobs luxuriantly. The
trucks are gone. . . . There is a $20 bill beside each plate."

It wasn't long before intellectuals, journalists, and sociologists
began to follow the tourist and the emigrant onto the highway. The
WPA, the program that built 651,087 miles of road, also celebrated
American road culture. The WPA's design index and building index
looked back at such elements of functional tradition as Shaker furniture, quilts, bridges, buildings. It was an agency of stocktaking

in a period of national stocktaking. It created, said one admirer, "A sort of road map for the cultural rediscovery of America from within."

In the course of its elaborate documentation of this culture—its American Design Index and Historical American Buildings Survey—it created a series of Guides to the states, to various cities, and even to roads—including Route 1. There are 378 of these remarkable books and pamphlets. Most are based on pedestrian tours in the cities and road tours out of them. They were at their best cataloging roadside oddities. This section is from a tour along Route 1 in northern Florida:

> The highway presents an interesting study of American roadside advertising. There are signs that turn like windmills; startling signs that resemble crashed airplanes; signs with glass lettering which blaze forth at night.... They extol the virtues of ice creams, shoe creams, cold creams; proclaim the advantages of new cars and used cars; tell of 24-hour towing and ambulance service, Georgia pecans, Florida fruit and fruit juices, honey, soft drinks, and furniture...

But the writer was also more conscious of the ironies of the roadside than the average traveler. Amid descriptions of the retirement home of Harriet Beecher Stowe, Indian mounds, and the Florida land turtle, known locally as the "gopher," we are told that:

> Along the highway, all but lost among blatant neon lights flashing "Whiskey" and "Dance and Dine" are crudely daubed warnings erected by itinerant evangelists, announcing that "Jesus is coming soon," or exhorting the traveler to "prepare to meet thy God."

The WPA sensibility picked up on the qualities of specific regions. Most recreational trips, naturally, were not cross-country but within regions. New England, the southern Appalachians, Florida, the Pennsylvania Dutch country, and of course the various sections of the West were strengthened as identities in the popular mind by auto travel.

The Guide to U.S. 1 supplemented its chapters with a guide to the culinary specialties of various regions: New England clam chowder, apple pandowdy in Pennsylvania, ham and fried okra in the Carolinas. Roadside restaurants began boasting these regional specialties.

The Guides reflected an ideology that, while not embarrassed, was barely conscious of itself—the references to labor and racial abuses and to the work of New Deal agencies in each place seemed merely fair play and the American way. The collective and anonymous nature of their authorship had to do with this. But to those outside the New Deal, the political bent was clear: South Dakota refused to distribute its Guide.

The Guides sent photographers and writers out along the roads of America to find the true state of things, the real interests of the people. Private enterprise—in the form of magazines like *Fortune*, which could hardly be said to share the ideology of the New Deal, shared a similar interest and launched projects paralleling the government efforts.

The attitude toward the culture of the road in these business quarters was one of bemusement, slight condescension, but above all fascination. The business press, Warren Belasco has noted, was particularly obsessed with the entrepreneurial strength of roadside culture, which continued to prosper even in the midst of the Depression.

In 1934, *Fortune* assigned James Agee, then a young staffer a couple of years out of Harvard, who had spent a recent summer hitchhiking across the country, to do an article eventually called "The Great American Roadside." Agee was captured by the American restiveness shown in roadside culture,

> a restiveness unlike any that any race before has known. . . . We are restive entirely for the sake of restiveness. Whatever we may think, we move for no better reason than for the plain unvarnished hell of it.

> So God made the American restive. The American in turn and in due time got into the automobile and found it good. The War exasperated his restiveness and the twenties made him rich and more

restive still and he found the automobile not merely good but better and better. It was good because continually it satisfied and at the same time greatly sharpened his hunger for movement: which is very probably the profoundest and most compelling of American racial hungers. The fact is that the automobile became a hypnosis. . . .

Agee was fascinated by the economics of the motor court and doghouse restaurant owner, the $630 million size of the aggregate formed from such small personal enterprises. He even contemplated writing a whole book on the subject. When forced, like most commentators on the culture of the roadside, to make a listing, he turned it into a kind of mock-epic catalog. He was amazed by the variety of tourist attractions: the home of Lew Wallace, Civil War general and author of *Ben-Hur;* the World's Biggest Sweet Peas, in Orlando, Florida; the Log Cabin Farms motor cabins on U.S. Route 22 near Armonk, New York; Freda Farms ice cream stand near Berlin, Connecticut, shaped like three cardboard ice cream containers.

He described

> . . . such staples of American *turismo* as rag and rubber animals, little jig-sawed animals to set up on the lawn, sponge rock for the rock garden, slips of boxwood from Virginia estates, carved coconuts, birch-bark canoes, windmills, foxtails for the radiator cap, redwood bark, snake-skin belts, ashtrays made of shell, pennants, baby alligators, little crates of kumquats, Civil War bullets (minnie balls), stalactites, balsam cushions, and turtles carved with the name of the young blade's girl.

Faced with the variety of the roadside, the journalist could do little more than catalog. Such cataloging could become almost epic, but it required the attitude of the outsider, the alien, to appreciate it. It required wonder and perhaps a bit of contemptuous amusement.

> "I needed a certain exhilarating milieu. Nothing is more exhilarating than philistine vulgarity."

> —Vladimir Nabokov
> "On a Book Entitled *Lolita*"

In *Lolita*, road culture becomes self-conscious, ironic. The book presents a view of the road and of the whole vast country that the roads unfold as a realm for individual projection, for solipsistic recreation, whose miscellany is organized only by the tortured mind of the narrator/hero. The whole landscape becomes a mock-romantic register for Humbert's individual emotions. "How many small dead-of-night towns I have seen. Neon lights flickered twice slower than my heart." His fear takes the form of a huge tractor-trailer, and elsewhere that of a mysterious "Aztec red" sedan.

Lolita also provides one of the best documents we have of the American road before the advent of the superhighways. It is filled with a series of mock-epic catalogues, a takeoff of the AAA Guides Humbert uses to plot the trip: "all those Sunset Motels, U-Beam Cottages, Hillcrest Courts, Pine View Courts, Mountain View Courts, Skyline Courts, Park Plaza Courts, Green Acres, Mac's Courts."

In what can be read as a broad and perverse parody of the devices millions of parents have used to keep their kids quiet in the back seat, Humbert uses roadside attractions to help keep Lolita in his thrall, setting up each of them, with the aid of roadside signs indicating their distance, as one more putative goal of the endless journey:

> Every morning during our yearlong travels I had to devise some expectation, some special point in space and time for her to look forward to... a lighthouse in Virginia, a natural cave in Arkansas converted to a café, a collection of guns and violins somewhere in Oklahoma, a replica of the Grotto of Lourdes in Louisiana, shabby photographs of the bonanza mining period in the local museum of a Rocky Mountain resort, anything whatsoever—but it had to be there, in front of us, like a fixed star, although as likely as not Lo would feign gagging as soon as we got to it.

He notes the cute toilet signs "Guys-Gals, John-Jane, Jack-Jill, and even Bucks-Does." The observation, for all the madness of the narrator, is Nabokov's observation: detailed and almost scientific, as befits a professional lepidopterist who recorded most of it during

47,000 miles of summer travel during the late forties and early fifties, in pursuit of butterflies.

Nabokov himself never learned to drive. His wife was at the wheel; his was the view from the passenger seat. What the driver and the passenger see are phenomenologically different.

Lolita is made up of a catalog of observations. It is not concerned with the flow of the road, the dynamic of movement that is Jack Kerouac's concern. *On the Road*, by contrast, is a continuum of discoveries and departures. Kerouac writes with his head turned toward the landscape disappearing behind him. His book is an essay on the phenomenology of pure driving, of encounter and abandonment. Humbert is a cool and distant, if cynical and desperate, observer who uses precision and irony to keep his emotions at bay. In *On the Road* the emotion overflows in a stream of present participles.

In Kerouac the pretense of a goal is dropped almost completely. His heroes are driving nowhere in particular—just "to the Coast" or back—which is the same as somewhere in general. They are existential drivers. The driver without a destination is kin to the rebel without a cause.

> What is that feeling when you're driving away from people and they recede on the plain until you see their specks dispersing?—it's the too-huge world vaulting us, and it's good-by. But we lean forward to the next crazy venture beneath the skies.

Kerouac's heroes always drive well, but too fast, taking some imagined existential risk for discovery. Speed and danger are part of the thrill of Kerouac's road, the heat and pressure in which the driver can forget himself and concentrate only on the beat of the wheels.

The superhighways gradually domesticated the wilderness of the old roads, settling and closing the highway frontier, eliminating the possibilities of escape. But the creation of the superhighway also strengthened the American fascination with the roadside, adding to its exclamatory, cataloging, wondering tone the elegiac haze of

nostalgia. It put an end to the glory days of the pure open road—
"In the rosy dawn, on a bridge over a superhighway, we said,
goodbye," wrote Jack Kerouac—but it also transformed the old
roads into myth. The back roads became a new legend of the "heart
of America."

In fact, of course, the back roads were simply the residue of an
old America, changed and gone forever. The real America *was* the
Interstate and the franchise. John Steinbeck, returning to the road
at the end of the fifties, found it a very different place from the
one the Joads had known. The old conviviality of the Depression
roadside was gone. In *Travels with Charley*, he lamented the effects
of prosperity and the superhighways. It was now possible to cross
the country without seeing it.

Steinbeck had become one of the millions of "tin can tourists"
whose wanderings had in part been inspired by *The Grapes of Wrath*.
Originally, the tin can tourists had been motorists who stopped to
eat out of tin cans. With the coming of the Airstream and Winne-
bago, the recreational vehicle and the pickup camper, it meant those
who traveled in "tin cans"—metal trailers, the first of which was
produced by the aptly named Covered Wagon Company. By 1935,
the WPA Guide to Florida reported, there were already 35,000 of
them. By the fifties there were millions of RV monarchs, in their
Airstreams and Winnebagos and Coachmen, straight out of the
heart of the country, the world's RV capital, Elkhart, Indiana.
They bore nicknames painted on them like boats—new prairie
schooners—and stickers of all the parks and states and towns they
had visited and decals of leaping trout, as if in token of each fishing
trip, like the bombs the ground crews painted on bomber fuselages
to mark each of their missions.

It was ironic that voyagers in recreation vehicles and campers
constantly complained about the changes the superhighways had
wrought, for in these space ships and land yachts of the highway,
they were the most cut off from the country of all, with little need
to stop at restaurants or motels.

The car had always provided a sort of protective visual bubble.
The automobile industry had virtually created cheap plate glass.

Henry Ford bought Libby-Owens, the firm with the first efficient, steady-stream method for rolling the glass, and at one point it was estimated that more than half of all plate glass went into automobiles.

The generous use of glass turned every vehicle into a sort of mobile greenhouse. And to the serious traveler this made his experience almost transcendent: his vehicle aspired to the status of Emerson's Transparent Eyeball, with 360 degrees of view.

The recreation vehicle added to the automobile an entire portable dwelling, with kitchen, bedroom, and living room. Its inhabitants rarely needed to stop at all, and when they did it was at a "camp" with "full hookups"—power and water for the onboard equipment.

Charles Kuralt, the television correspondent, repaired to an RV to search out his offbeat reports of offbeat places. He could travel along 66 to find the Cadillac Ranch, near Amarillo, and help make it famous. Kuralt's journeys, like Steinbeck's, were part of a constant effort to reassure us that the country still possessed the diversity of the past that standardization, while regularly lamented in such accounts, had not completely conquered America.

In William Least Heat Moon's tremendously popular *Blue Highways* the romance of the backroads is not just confined to the landscape, but is projected as a series of instant and easily available epiphanies. At practically every turn, waitresses and guides, old men sitting beside streams and the proprietors of gas stations, turn into prophets of the real America, as they had figured in the road writing of the twenties and thirties.

Road nostalgia was carried to full flower in the press and television reports on the end of 66. These became virtually a genre in themselves: the Lament for Route 66. Sixty-six, we were told, was being bypassed. The road of the Okies and of the migration to California, with its lovely little Ma-and-Pa diners, its motor courts and teepee motels, was being killed by the cold, impersonal Interstates. Any day now, went the stories' "news peg," the last link would be finished and 66 would be gone.

One group of mythologizers of 66 located the mother of country

244 Four: Gone to Look for America

singer Merle Haggard, whose family made the trek from Oklahoma to Bakersfield shortly before he was born. Interviewed in the seventies, she sounded like one of Steinbeck's characters. She told how the family car overheated in the Mojave. A boy on a bicycle stopped to give them a quart of water. "Everybody'd been treating us like trash, and I told this boy, 'I'm glad to see there's still some decent folks left in this world.'"

A typical report was presented in 1984 by Bob Dotson, a correspondent for NBC's *Today* show, who bubbled that Route 66 "was a thoroughfare for freedom, beat across the wilderness, a timewarped ark upon which so many of us determined the dimensions of our American dream. Sixty-six was for most a yellow brick road, a journey important for what we would find." He added the ritual denunciation of the Interstate, which, with the bypassing of Williams, Arizona, had completed the killing of 66. "The Interstate leaves little history. Everything is too fast. On 66, there was time, time to sleep in cement teepees and time to read messages on rusty fences, and laugh."

Almost all the eulogists of 66 stopped by the Cadillac Ranch, near Amarillo. This odd pop sculpture was the definitive monument to the myth that Route 66 had become. Located in a wheat field beside 66, the Cadillac Ranch was a set of ten Caddies embedded halfway into the ground at an angle and arranged in order of their model years. The Cadillac Ranch was commissioned by an eccentric Texas millionaire and art collector, Stanley Marsh 3, and created by Ant Farm, a sort of counter-culture art collective whose members had spent some weeks touring and videotaping the country in a "media van" in a reprise of the project of Kesey's Magic Bus.

The Cadillac Ranch, said Doug Michels, one of its creators, was inspired by the wheat field where it was built, and visions of tailfins dancing among the wheat like the fins of dolphins amid waves. It was a democratic sort of monument: its owner made no effort to protect it from the graffiti, bullet holes, and eventually the complete coat of red paint that visitors added to it.

Collecting the Caddies, said one of the principals of Ant Farm, "was a white trash dream." Every Okie aspired to the Cadillacs that Steinbeck described whizzing by their old jalopies in *The Grapes of Wrath*. Elvis Presley was famous for dropping in on Cadillac dealerships, selecting some couple wistfully and hopelessly admiring the cars through the windows and immediately bestowing on them the car of their dreams.

The Cadillac Ranch could easily be read as a monument to the end of the gas guzzler and, by extension, of the endless and speed-limitless road. El Dorado was no longer a golden goal—real or imaginary—at the end of the trail; it was the name of a vehicle. But in fact, the Ranch was a monument to roadside camp, to the "exhilarating vulgarity" of road culture grown self-conscious.

Clark Gable and Claudette Colbert (far left) on the road and on the lam, in Frank Capra's comedy It Happened One Night, *1934. In the same year, Bonnie and Clyde met their end at a Louisiana roadblock, inspiring road films whose tragic couples would flee their doom down dark highways. (COLUMBIA PICTURES)*

17: Escape Routes, or They Drive by Night

Whither goest thou, America, in thy shiny car in the night?

—*Jack Kerouac*, On the Road

Clyde Barrow and Bonnie Parker, the most romantic of modern outlaws, met their end in a hail of bullets at a roadblock near Gibland, Louisiana, in 1934. Clyde was driving without his shoes; Bonnie was eating a sandwich.

They knew their deaths were inevitable, but the road, the backroads of Texas and Oklahoma along which they had led law enforcement officials a merry chase for years, offered them temporary escape. In the process, with letters and photographs sent to newspapers and wire services, they built up their legend and the legend of the outlaw on the road. Bonnie posed with a pistol and a cigar, her leg hitched up on the bumper of the couple's Ford. Clyde went so far as to write to Henry Ford, praising the virtues of his V-8 as a getaway car.

It was a legend that was to become central to American *film noir* and later to American rock 'n' roll, a legend of the romantic outlaw that Kerouac and others played out on the road. Three years after Bonnie and Clyde died, a story very much like theirs was recounted in Fritz Lang's *You Only Live Once*, a film full of the romance of the fugitive couple, fleeing along back roads into the American night, shot from a dramatic angle. Lang had been a pioneer at capturing the sensations of the road at night in his German movies, as early as *Spies* (1927), where motorcycle riders dash along in an early chase scene, shot from extreme angles. Translated to the American highway landscape, his techniques set the pattern for a whole series of *films noirs*, tales of romantic Robin Hoods fleeing along blind highways.

The myth behind such films is that, on the road, escape is still possible. It is outlaw turf, where no one is tied to a particular identity and in whose darkness everyone is a stranger. *Film noir* lent the road some of the air of danger that it had occupied long before the automobile, when brigands and wild animals lay in wait for the uncautious traveler. For the criminal, the road offered a new area for escape. J. Edgar Hoover may not have been so far wrong when, in an influential article in *American Magazine* in 1940 he called tourist camps "camps of crime." The car and road not only made it possible for criminals to flee more quickly from state to state than horsedrawn vehicles and more surreptitiously than rail or steamship, but offered a series of hideouts along its sides.

The legend was still alive thirty years later, when Arthur Penn's *Bonnie and Clyde*, with its careful attention to the details of the gas stations the pair held up and the motor courts where they hid out, and, of course, its slow-motion death scene, became a sensation.

For American filmmakers in particular, the chase, whether the short-term chase scene or the extended flight of fugitives, became a stock in trade. Much of *Lolita* is one long chase scene—although who is being chased and why is mysterious. This is the aspect of the book emphasized in Stanley Kubrick's film—with its inconsonant English backdrops—and dozens of other American films have turned on plots of chase or flight on the road: *Sugarland Express*, *Alice Doesn't Live Here Any More*, *Badlands*, *Five Easy Pieces*, *The Getaway*, *Starman*, *Easy Rider*, *Convoy*, the whole genre of "Burt Reynolds films." The chase could be comic, as in Frank Capra's classic *It Happened One Night* (1934) in which Clark Gable and Claudette Colbert take flight by hitchhiking and public bus, among other means.

In the crudest films and television shows the chase scene becomes a sort of pornography of the road, where speed limits and rules of the road are defied equally by cop and robber and forbidden fantasies are realized. The car-driver unit in these scenes possesses special dexterity and unheard-of aerial skills. It miraculously avoids unavoidable collisions, accomplishes hair-raising leaps over opening draw bridges, and always finds some happy, impromptu ramp from which to leap over other vehicles entirely. The frugal director can stretch and stitch the same footage into dozens of such scenes. This sort of acrobatic, kinky driving appears at regular, predictable intervals, like sex scenes in pornographic novels. In many such adventures, the road is a scene for chases that are stitched together by plot that is conventional and often merely nominal.

The road here is sometimes a physical correlative of the twists and turns of plot, but more often a substitute for it. It also casts an aura of strangeness and otherness around the roadside. The landscape looks different to the outlaw, the traveler with a serious purpose, the alien—it regains a certain freshness brought on by fear and suspicion and the incidental feel of all objects removed from their main purpose. Such an attitude corresponds to the false casualness of the movie camera, which packs its images with details that only seem accidental.

The implicit linkage this movie myth made between violence and the road touched the larger dangers every driver faced on the highway. The death rate was the price Americans paid for the freedom of the highway. The individual seemed to be his own master on the highway, but he was also at the mercy of the competence of other drivers, vulnerable to collective danger: the nut who veers across the centerline or the stalled truck.

Death by auto is the most absurd and existential of deaths. It can be courted by speed, recklessness, or drunkenness, or it can come by crude chance: to some extent, any driver is at the mercy of all others; he gambles every time he goes on the road. The absurdity of the myth of the driver as his own master is made clear when one or two cars block an entire road: it is documented in the traffic backup in Godard's famous long tracking shot in *Weekend* or in the chain reaction collision in Robert Altman's *Nashville*. It was perhaps a collective version of this myth that enabled American society to come to terms with 50,000 traffic deaths annually.

It was hard not to suspect that Americans somehow enjoyed the edge of danger the road provided, the fact that in many cases death lay only a few inches away, ready to be tapped by a sudden jerk of the wheel. On the road people lived out some version of the road myths they knew from film.

The open road has always ministered to the American flight from self. To drive without purpose—to "cruise"—is the central trope not ony of Kerouac but of a hundred popular songs, in country music or rock 'n' roll. Just driving, without goal or purpose, surrendering the mind totally to the mechanical functions of steering wheel and gas pedal, figures in such songs as a solace.

Carried further, such driving is an expression of self-destructiveness. Speed begins as a flight from identity—an attempt to outrun the past—and ends in the crack-up. Neither physiology nor psychology had ever quite pinned down the source of speed's appeal. Did it come from the mental rush that followed from the rush of passing images? Or was it a consequence of acceleration, forcing the blood back in the body, a thrill dependent not simply on speed but on a constant increase in speed?

As time went on, the film version of escape by highway became more and more stylized. The backroads were fading as a locale for escapes. By 1949, in Raoul Walsh's *White Heat*, Jimmy Cagney and gang, after a shoot-out at a motor court, elude their pursuers by turning into a drive-in movie theater—a humorous escape into the new wilderness of roadside sprawl.

In the fifties, the romantic criminal myth, at least as old as Schiller (Lolita's married name), that figured in Lang or in Nicholas

Ray's *They Live by Night*, was succeeded by the drag racing myths of his *Rebel Without a Cause* or the motorcycle fantasies of Marlon Brando in *The Wild One*. The romantic rebel had been replaced by rebels whose escape was simply driving, drivers without destination.

From film the myth of escape on the road found a refuge in the tropes of popular music. The blues had long offered a musical romance of escape—up Route 61, from the Delta to northern cities—and the first flowering of its heir, rock 'n' roll, captured the rhythms of the road. The sound of wheels and the throb of the engine provided the backbeat for such classics as Chuck Berry's "Maybellene" and "No Particular Place To Go," songs the teenagers of the baby boom listened to on their car radios while pursuing the quintessential recreation of the fifties, cruisin', driving aimlessly about to see and be seen, with stops at the drive-in, "the passion pit." In Bruce Springsteen's songs, full of images from *film noir* late shows, the escape has become little more than a high-speed drive over the night roads of New Jersey, the physical correlative of a hopeless dream of escape from the banality of everyday life.

In country and western music, too, with its recurrent images of restlessness, "rambling," and flight, the myth of the road as escape flourished. The "lonesome fugitive" Merle Haggard sang about claimed the highway as his home, like Steinbeck's Okies. Haggard's family had taken Route 66 out of the Dust Bowl before he was born, and later, as a fugitive from justice, he had hitchhiked through Arizona back along its route.

For the most part, however, the road was a state of mind for country music, a view abetted by the roadshow nature of most of the business: Willie Nelson said that he wrote many of his songs literally on the road, driving between show dates, and Hank Williams entered legend by dying inside his Cadillac, en route to a performance.

While the images of road films danced through the head of the road traveler, the songs of the road played on his car radio. But the vacationer's escape was temporary, a set of parentheses in the normal flow of life. All that he brought back was a set of memories, echoes, roadside souvenirs—and photographs.

The ubiquitous gas station, captured in Reedsville,
West Virginia, by Walker Evans. The building and
the signs are all part of a single collage of roadside
imagery that has become universally familiar.
(LIBRARY OF CONGRESS)

18: The Mobile Eye: The Camera on the Road

When eighteenth-century French aristocrats strolled through the countryside, they viewed the landscape through an odd device called a "Claude glass"—a framed piece of smoked glass used to capture picturesque views like those in Claude Lorrain's paintings. For the modern highway traveler, the Claude glass is the tinted windshield.

Perhaps that is why photography, whose framing is a descendant of the Claude glass, has shown itself to be a medium particularly well suited to recording the culture of the road, to capturing picturesque cross sections of the flow of road life and the play of roadside imagery. Its single, personal viewpoint, its seizure of moments from the flow, is perhaps the only way to bring an immobile order to the visual experience of the highway.

Early in the century, what interested the photographer was the speed of the car by contrast with the fixity of road and roadside. Lartigue and others made pictures of blurred racing cars—pictures that suggest the shimmering auto images of the Futurists and Marinetti's breathless declaration that "a racing car is as lovely as the Nike of Samothrace." Americans were more intersted in the signs, buildings, and people along the wayside.

The roadside "picturesque" is complicated—ironic, personally composed, created by juxtaposition. Photographers of the roadside can pick up ironies as cheaply as roadside knickknacks or manipulate a world of stage flats for the sake of drama. They can make photography turn signs advertising food or car parts into literary messages.

Photography can show the roadside at its best and worst. It is easy, for instance, to create a picture establishing the ugliness of the strip by exaggerating its chaos with a deep-focus shot, pulling a mile's worth of signs and poles together into a ghastly spaghetti. But it is also possible to carefully compose this same scene into a whole, to render permanent something like one of those brief and sudden views that give pleasure along the road.

The best road photographs resemble film stills—in their sense of casualness (an implicit bid for us to release them from the more formal rules of composition), in the inexplicability of their situations, in their technical imperfections, in their sense of being "establishing" shots for silent dramas that we can only imagine unfolding.

The photographers who accompanied the railroad surveyors and builders produced the first photographs of the great west. The pictures taken by men like Carleton Watkins, George Barnard, Eadweard Muybridge and others were godlike, definitive, full of an impersonality that evidenced the photographer's sense of responsibility, of being first, of being a delegate of his viewers, an explorer with a report to bring back.

Their pictures possessed the same sharply detailed power as photographs of the moon—and of battlefields. As generals and military organization moved from the Civil War camps to the rail projects, so some battlefield photographers, like Alexander Gardner and George Barnard moved west as well.

Highway building produced few such images. Photographers of the roadside wrapped their images in the apparent casualness of the tourist snapshot, the souvenir, the things the vacationer brings back from the road. For photographers, the roadside landscape provides a kit of parts from which to compose personal views. The parts of this kit lent themselves to formal composition, with flat pieces to "acknowledge the flatness of the medium," and elements that push and pull in depth to keep the image in suspension. They offered extreme versions of the order that the driver or viewer "composes into" the roadside.

Walker Evans helped transform the tradition of the street photographer—the tradition of Atget and others—into an American one of the road photographer. Evans had begun as a street photographer, fascinated with signs: at Coney Island, he framed sign-boards at Luna Park into something like a Stuart Davis abstract painting. On the streets of Havana, he focused on ads of exotic script on whitewashed walls. He captured the dangling legs and canes in front of a New York prosthetics shop, the carefully hand-scripted menus of a restaurant, the crudely painted sign of a shop, with the signpainter's name and address—a miniature advertisement in itself—readable in the lower righthand corner. But his travels for *Fortune* and for the Farm Security Administration produced one of the most telling collections ever assembled of images of roadside signs.

Evans shared the enthusiasm of his friend and collaborator James Agee for the do-it-yourself creativity of the roadside. In Evans's roadside, planes of imagery and lettering—often crude and home-made, intensely human—appear suddenly, interrupting the long perspective. They replay the tricks of distance and scale of early roadside architecture in order to suspend the words and letters in space, floating on their own.

In a gas station in Reedsville, West Virginia, that Evans photographed in 1936, the whole building is itself a giant sign, its front and side walls proudly painted with its name in huge letters and the standardized signs for three different brands of motor oil packed in under its porch along with three crudely handlettered warnings within ten feet: "Terms: Cash."

The station owner doesn't want to have to get into a discussion about credit: he wants to remain friendly, just folks. (The *no credit* sign is a versatile genre of American sign, from the familiar "In God We Trust; All Others Pay Cash" to such labored individual efforts as—this seen in Louisiana—"Credit manager Helen Waite. If you want credit, go to Helen Waite.")

The station owner is also boastful of the variety of products he sells, established national products, the sophisticated graphics say.

This photograph, however, is also a parable of the visual structure of the roadside environment. In Evans's framing, the gas station seems to grow out of a pole in the foreground, like just another panel on the post. Letters, having no inherent size, scramble the signals of perspective.

Often the billboards in Evans's photographs are confused with the reality around them: grass on an ad for a furniture store picks up from the grass in front of the sign; the palm trees cut out of the edge of a partially painted billboard for a Florida hotel are parodied by the pine trees behind it. Evans was aware of the pleasure this confusion of realities gave to the traveling public, the fantastic visual element that it lent even to mundane products and sales pitches.

Evans adored the flat front, the corrugated tin of a contractor's shed, the signs on the little shack, a transplanted Western town front, with the signs for Clabbergirl Baking Powder, patent medicines called 666, Grove's Bromo Quinine and Grove's Chill Tonic, and, of course, Coke. He liked the handmade sign, the amateur's price list or apotheosis of fruits and vegetables, carefully and tortuously painted, like a child making his first block letters.

Evans collected these signs in his photographs the way he collected postcards—the fundamental souvenirs of the road: giant jackalopes and apples or potatoes collaged onto railroad flatcars, scenic vistas captured from the one best spot, on the one best day of the year, desert scenes with the colors hyped into peyote fever, made-up tourist attractions and dull motels, ready to be marked "This was our room." And by the end of his life, when he was taking fewer and fewer photographs, Evans was collecting the signs themselves.

Evans's work was far from typical of the pictures of most of the Farm Security Administration photographers. His interests were more formal than social. Most of the FSA photographers, while their work seems to provide a counterpart to the travelogues and tours of the WPA Guides, were concerned with social and political reportage. Here too, the road—Route 66 and all it stood for—was the organizing element.

Dorothea Lange went on the road to look for Americans on the road who were looking for America: looking in California, specifically, for the economic promise that had failed them in the Dust Bowl. They and she found the failure of the promise of the frontier—the old promise that if you just move west, there will be land and where there is land there is prosperity.

The road is the physical embodiment of this illusory promise.

The cover of *American Exodus*, the 1939 volume produced in collaboration with Paul Taylor, shows a covered wagon; inside, a view of an automobile makes the point unsubtly: "Covered Wagon, 1939 style." Early in the book there is an archival photograph of the 1893 land rush in the Cherokee Strip of Oklahoma—a view, presumably, of how the Okies got to Oklahoma in the first place, how the promise was first made.

The book aims to trace the economic and social process that created the Dust Bowl. It assumes a documentary stance—just let the pictures speak for themselves—then tips its hand by supplementing "factual" captions with quotations, unattributed but by implication the words of the people in the pictures.

Chapter titles further undermine the strictly factual air. "Human Erosion" is the title of one chapter. *American Exodus* is a virtual companion volume to *The Grapes of Wrath*, also published in 1939. Pictures with captions like "Breakfast beside U.S. 99" show the communality of the road Steinbeck makes much of. "Us people has got to stick together to get by these hard times."

In addition to the photographs of the people themselves, to the views of old cars packed with furniture and livestock—a cleverly penned goat in one picture rides the running board—there are photographs of the roads themselves.

Roads are seen both as part of the problem—"Highways are part of the process of mechanization [of agriculture]" reads one caption—and also the focus of the elusive hope of its solution. "Highway to the West. 'They keep the road hot a goin' and a comin'.' 'They've just got roamin' in their head,'" reads the caption to a photograph of U.S. 54 in southern New Mexico.

Just as many of Evans's photographs of the interiors of sharecropper houses surpass his portraits of their inhabitants in force, it is in the photographs of roads in the desert, where distance and desolation unfold to the horizon, that Lange's camera achieves a documentary force more lasting than in the frequently sentimental photographs of the Okies themselves.

Evans had a more expressionistic successor in Robert Frank, who captured moments of roadside juxtaposition in *The Americans*. Frank's

images, taken in months of driving around the country in 1955 and
1956, courtesy of a Guggenheim Foundation grant, contrast the
incidental to the constant structure of the road itself, even when it
is offstage. Formal elements—the coffin and juke box, a shrouded
corpse in the Midwest and a carefully tarpaulined automobile in
California—echo each other as one turns pages in *The Americans*, as
visual memories echo one another as you move down the road.
Frank's motto might have been a paraphrase of Heraclitus: "You
never photograph the same road twice."

The Swiss-born Frank was a sort of photographic de Tocqueville:
fascinated by America and appreciative, but also critical and cautionary, and possessed of the sharpness of observation that, especially in America, comes more easily to the outsider.

In his pictures there are ironies whose heavy-handedness is disguised by the determined anti-composition of the grainy, gray images themselves: the raised cross of the statue of a monk, who might
have helped blaze the Camino Real, over the buildings of the strip;
the giant SAVE sign and the gas station. But there is also pure
feeling and mystery that contribute, in *The Americans*, to a sense of
a complete if personal vision: a covered corpse and huddled mourn-
ers in the snow beside Route 66; a carefully covered car beneath
palm trees, under a bright California sky, somewhere beyond the
end of 66.

If Evans had his equivalent in cataloging the roadside in Agee,
Frank's literary counterpart was the adjectival hysteria, the stream
of fevered consciousness that took its rhythm from the road, of Jack
Kerouac, who wrote, in his introduction to *The Americans*, "Mad-
road driving men ahead—the mad road, lonely, leading around the
bend into the opening of space towards the horizon Wasatch snows
promised us in the vision of the west, spine heights at the world's
end, coast of blue Pacific starry night—nobone half-banana moons
sloping into the tangled night sky, the torments of great formations
in mist, the huddled invisible insect in the car racing onward..."

While Evans and Agee went on the road to look for things particularly American—and if often transcendent, then still transcendent in the American mode—Frank and Kerouac were concerned
with a personal sort of discovery and composition on the road.

They are drunk on space, high on speed, racing back and forth cross-country not for the sake of something to be found but for the sake of the looking itself. "That's not writing, it's typing," was Truman Capote's famous putdown of Kerouac, and for both Frank and Kerouac the point was not getting somewhere, it was riding.

They were cowboys of the open road, tossing out the "lassos" of perception of which James had spoken with wild west abandon, loners recreating the roadside in their own image—a combination of observation and chance.

Evans has other heirs: in Lewis Baltz, whose pictures of industrial parks and condo developments along the highway take the same pleasure in stagey, textured planes as do Evans's views of rural stores and cafés; in Robert Adams's views of Denver sprawl; in Stephen Shore's local-color compositions of strip structure, vegetable stands or Holiday Inn place settings; in Lee Friedlander's roadside views, sometimes including his own face seen in a vehicle's side mirror.

Stephen Shore most successfully adapted Evans's techniques to cool, color views of the new franchise strips. Shore grew up in New York City, and began his road photography in 1972, when he rode with a friend from New York, down to Route 66 and Amarillo. Like Nabokov, he could not drive—he saw the landscape, he said, from "the passenger window." His first trip through America "was a shock. I would be in a flat nowhere place of the earth and every now and then I would walk outside or be driving down a road and the light would hit something and for a few minutes the place would be transformed."

One of Frank's admirers, a photographer named Reed Estabrook, took his project to something like its ultimate extension, turning Frank's formal devices for expressing the juxtaposition of continuity and anecdote on the road into mechanical ones—"auto-matic" photography. Estabrook systematically took photographs from the back of a van as he drove from San Francisco to New York, along the 2,900 miles of Interstate 80.

Estabrook considered himself, he said, more a conceptual artist who uses a camera than a traditional photographer, and he looked at the I-80 photographs as one single work, which he called *North American Cross Section*.

Estabrook had taken the format of Ed Ruscha's *Every Building on the Sunset Strip*—a kind of deadpan realtor's collector of facades— and extended it to national scale. He compared his project to one of Sol Lewitt's "generative drawings." He called it "a photo-mechanical device to explore a perceptual concept—time and space."

Estabrook's big picture was also a sort of parody of vacation out-the-window shots. He even had his rolls of film—250 exposures apiece—processed at a fast film outlet, one of those stores where, on machines seen through the windows, people's snapshots go rolling by. He ended up with more than ten-thousand square feet of film, a strip 3.3 miles long if assembled—Estabrook has never been able to get the funds to display the entire project.

The premise of his piece shares the same claim of the cross-country vacationer to have "seen the country." What you can see from the highway is extremely limited. You don't for the most part see countryside at all, you see roadside. But having driven across the country, you can say you have "seen America."

Estabrook's equipment consisted of a Nikon set at f2 and embed-ded in the side of a Dodge van, six-feet-two-inches above the ground. Keeping a steady speed of around fifty miles an hour, Estabrook triggered the system manually, methodically, almost rhythmically, having figured the alignment between what he saw out his window and what the camera saw. He tried to link the photographs by means of what he called "the distinctive feature of the landscape," whether manmade or natural. In densely built-up areas, where the dominant feature was a roadside restaurant or used car lot, his program forced him to take many more pictures for the same dis-tance traveled than in the open spaces, where a distant mesa dom-inated the landscape for miles.

The result is a series of views with much in common, much in disjunction. Weather seeps in, trailing or catching up according to the wind. Incidents occur in the frames: clouds like buffaloes or schools of minnows, pennants flapping from the gas station lots, shadowy figures caught entering convenience stores.

Estabrook passed near where the Donner Party died—an area of crumbling rocky bank dotted with low firs. He photographed, almost incidentally, in passing, a great rock formation near Echo Canyon that William H. Jackson made the centerpiece of one of his most majestic railroad survey photographs. An occasional passing vehicle falls into the view of some of the pictures, like a lumber truck with decoratively painted cab.

Once, in Nevada, Estabrook passed a couple driving nude in their Mustang convertible. They smiled and waved. But the camera missed them. Their appearance fell between frames and left no impression on the project. The couple was a token of all he missed in the interstices of the film, of the inexhaustibility of the road to even such methodical treatment—and the way time and incident creep in to alter the steady, unchanging forms the roadbuilders envisioned.

But in another sense Estabrook's work is a kind of model of I-80. It shows the road in a way no traveler would ever see it and it constructs continuities, harmonies, showing something formal and almost musical in its length.

Even when road photographs seem rigidly conceptual—in Estabrook's photos or in Ruscha's gas stations and parking lots—there seems to be a personal motive, an individual viewpoint that constructs an order for the roadside. Ruscha and Estabrook lend organization to the roadside by conceptual frameworks. Ruscha's gas stations form a typology of outlets miles apart. His parking lots reveal their shapes and orders—the fish skeleton of their painted strips—only from the air. His view of the Sunset Strip assembles the buildings facade-on, in a wholeness that the driver, proceeding along the boulevard and viewing them obliquely, never sees.

Estabrook's cross section is to be contrasted to the photographic collections accumulated by some state highway departments, who dispatch men with cameras in vehicles to document the condition of highways. These men spend months on the road, photographing each highway every couple of years or so, from the driver's perspective. They click the shutter automatically every hundred feet, providing a view of the route that engineers, sitting back in headquarters with a slide projector and remote control, can "drive" with an eye to visibility, and physical deterioration. Estabrook's method was perfectly suited to the space-time world of the Interstates.

Walker Evans had said that the postcards he purchased and the photographs he took were "almost the same thing." He wrote in *Fortune* in 1948: "A collector becomes exclusively conscious of a certain kind of object, falls in love with it, then pursues it.... It's compulsive and you can hardly stop. I think all artists are collectors of images."

Evans's model was to motivate a whole new generation of road photographers. With the Interstates and bypasses driving the individuality of the old roadside into extinction, they became its archivists, students and devotees, pursuing a mix of nostalgia and camp. Photography, for them, became a way of collecting, in mock-postcard form, the sites they discovered.

John Margolies grew up in the fifties in Connecticut, where he was fascinated with the "gasoline alley" of the old Berlin Turnpike. This early strip was the location of, among other roadside attractions, a famous ice cream stand referred to by H.-R. Hitchcock. It was composed of three oversized models of paper ice cream containers, (in a shape more often associated today with take-out Chinese food), deployed in Palladian style, with two smaller containers flanking a larger one. The best touch was the spoons leaning out of the chicken wire and stucco ice cream roofs.

Margolies was obsessed with old-time roadside culture. He spent days on the road photographing it and collecting postcards, maps and signs.

Margolies lives in a small apartment in Manhattan, filled with old gas station signs—the classic rearing red Mobil Pegasus, the Texaco star—and old road maps in wire racks on the wall, their covers bearing images of square-jawed, All-American gas station attendants with round and soundly blocked hats. There are enamel and tin soft-drink signs of the fifties roadside, advertising soda pops long since bubbled into obscurity. In one, for "Norka"—an Ohio brand, Akron spelled backwards—the highlights on the bottle reveal, on close inspection, the pattern of windows in the room where the painter sat.

Margolies lives for the road. He subsists on the occasional sale of his photographs—they decorated the trip West in Martin Scorsese's film, *Alice Doesn't Live Here Any More*—on books and magazine articles. From time to time, when he gets a grant from a foundation or endowment, he undertakes another trip. He rents the largest car he can find, from Hertz or Avis, and packs along a special pressed-wood backboard made by the Artex Company to assure his rest.

He tunes the radio to a Top 40 station. He photographs only on clear days: blue skies, dotted with occasional, harmless, token clouds, backdrop all his views—of Mammy's Cupboard, the Natchez gas station/café once photographed by Edward Weston, the crumbling "Trail" drive-in in Albuquerque, the red and yellow expanse of the Allentown, Arizona, Indian City, an old gas station in Fabens, Texas, an Egyptian miniature golf hole in Florida, Moody's Barbecue in Kingsland, Georgia, Hat 'n' Boots Gas in Seattle, with the restrooms in the giant boots and the cash register under the hat, and an ice cream cone sign in Verden, Oklahoma.

Margolies is one of a whole group of photographers and collectors of roadside camp. Allan Hess traced the origins of the first McDonald's—the pre-Ray Kroc, California McDonald's that the company prefers to forget. John Baeder photographed and painted diners and gas stations. J. J. C. Andrews collected photographs of odd roadside buildings, from the Noah's Ark in western Pennsylvania to the giant oranges of Florida. Charles Schoeneman retrieved the glass lamps from old gas station pumps, painted them up, and put them on sale to collectors.

Oddly, the superhighways that turned the older roadside culture into camp, with the patina of the bypassed and the aura of abandonment, also made possible the wide-ranging trips of the photographers and collectors. When he travels, John Margolies seeks out the Holiday Inns at the Interstate interchanges, shunning the motel restaurants in favor of fast food. When it rains he stays in the motel room, watching game shows until the clouds break.

Spanning the continent, Interstate 80 slices through Telephone Canyon east of Laramie, Wyoming. One of the busiest interstates, I-80 already needed major reconstruction by the late seventies. Like the American road, highway construction promised to go on forever. (WYOMING HIGHWAY DEPARTMENT)

19: Ends of
the Road

Space launches at Cape Canaveral had always drawn huge, au-
toborne crowds, but a record number turned out for the first launch
of the space shuttle, in April of 1981. Recreation vehicles and
pickups, Cadillacs and Toyotas, assembled from all over the country. It was the first major launch in nine years and it came just
when, with the passing of the gasoline shortages of the mid-seventies,
Americans were returning to the road in earnest.

For the odologist, Florida provides a neat summary of the history
of road culture. Roads of various types are squeezed together down
the peninsula like geological strata: the old military road, built when
Jefferson Davis was Secretary of War and the Seminoles were still
unpacified; Carl Fisher's Dixie Highway, by now widened and
overbuilt; U.S. 1 and the nearby, more wandering, A1A, classic
tourist strips with fountains of youth and gator farms, tin-can tourist
parks and drive-in churches; the Florida Turnpike, an early toll
superhighway; and I-95, almost next door.

Down all these highways the space tourists flowed. They settled
in by the inland waterway. They stopped to hook up at KOA
Kampgrounds. They perched aluminum chairs up high beside the
moon roofs of their recreation vehicles. They registered at the seedy
row of motels along Cocoa and Satellite beaches, remnants of the
first space boom, whose signs featured aquamarine amoeboid shapes
with flaking paint, and neon rockets with burned-out nose cones.

Beginning at two or three in the morning, the traffic backed up
on the freeway into the Space Center, which ran among primeval

swamps resembling the backdrops in illustrations of dinosaurs. The ruby chain of brake lights, flashing on and off in succession, registered an occasional brief forward shudder of the traffic.

Two whole lanes, the outgoing ones, were empty and the others full as the sun began to rise. First a van from a television station, with a little dish antenna on its roof, dashed across the grass median and headed down those lanes in the wrong direction. Then, as the sun came up and the time of the launch grew near, drivers began to lose patience. First a red Firebird spun across the median, leaving ruts and tossing back a divot, and ran down the empty lanes, to be followed by another and another car until, ten minutes later, a blue Air Force military police car came and the MPs diverted the stream back into its legal channel. It was proof of the fragility of the order and etiquette of the highway.

We had never imagined the road culture to be so delicate, so vulnerable to disruption. But when the gas lines arrived, the crisis of the highway seemed to put in immediate physical terms the other crises of national character: Watergate, the fall of South Vietnam, and the Iranian hostage crisis. The crisis of the highway stood as an expression of some larger national failing of energy and identity, a weakening of the national skeleton.

Roadbuilding was like the space program: it had long been presented to Americans as yet another reassurance that the frontier would remain open, that the horizon would continue to expand. Americans drove a trillion and a half miles a year—a good start on the way to the stars.

Now the analogy was being reversed. When NASA's bureaucrats came to the Capitol to justify the space shuttle to Congress, they not only talked about space as "the final frontier," "the endless frontier," but constantly compared its eventual economic and social benefits to those produced by the transcontinental railroad—and the Interstate highway program. They sold the space shuttle as "a space truck" that would open the road to the heavens.

This approach held particular appeal to Americans in the seventies, even if the old space program of the highway itself had begun to lose its luster. The energy crisis had raised a new fear, the fear of reaching the end of the road. To complete or abandon the road system, as falling gasoline revenues threatened, was another form of the closing of the frontier.

In the face of such threats, Americans turned back to their beginnings. The crisis drove them back into nostalgia for the highway past, into an appreciation of old roadside development (150,000 gas stations—almost half of the total—and thousands of drive-in theaters went under in the seventies), of lost highways that rose and fell with the hills, swooped and twisted with the curves, roads on which you could still feel the texture of the pavement in the seat of your pants, roads, wrote William Least Heat Moon in *Blue Highways*, where "a man could forget himself."

The novelty of the superhighway worn off, backroads grew more attractive. And the car began to seem less important than the road, perhaps because the American automobile industry had so badly failed to match the energy-conscious times. In the fifties and sixties the car companies had touted the ride and suspensions of their cars, as well as their brute power, as if they floated over pavement like aircraft over the landscape. A famous Ford ad featured a diamond cutter doing his delicate work in the backseat of a car going over potholed pavement.

But in the seventies and eighties, with power and size limited by fuel efficiency and pollution controls, the selling point was handling. Now Ford urged drivers to take its cars over the back roads. "Thunder roads," proclaimed one ad, decorated with photographs of a group of twisty, roller-coaster highways. "The engineers who designed these roads knew that they'd present a challenge. But they never expected that some would drive these roads in a Thunderbird for precisely that reason."

The crisis marked the completion of the road culture's loss of innocence, but for a while it also restored a bit of the camaraderie of the early auto days. There was something of the sense of wartime, of shared national deprivations and efforts. Suddenly the highways seemed like wilderness again. The dangers of finding gas stations closed or of being apprehended for exceeding the new fifty-five-mile-an-hour speed limit added a hint of adventure to driving—a man-made version of the dangers faced by the first autocampers.

It was significant that during this period the trucker became a hero. The trucker was a man who drove for a living, whose office was the highway. He had always been viewed as a kind of modern cowboy, a tough westerner: Humphrey Bogart in *They Drive by Night*.

But what appealed to the national mind now was that, especially with the crisis, the trucker was an independent man fighting the diminution of his freedom, oppressed by schedules as well as the shortages. Country music was full of songs about this sort of trucker, pining for the end of the road, but never settling in once he got there, in love, it seemed with distance and his own loneliness. He was, sang country musician Ronnie Millsap, a prisoner of the highway, imprisoned by the freedom of the road.

Ignoring the truth—that these were mostly soul-tired, Benzedrine-driven guys, that trucking was a big, tough business, subsidized by highway taxes, infiltrated by organized crime, putting larger and more dangerous vehicles on the road all the time (the largest, nicknamed Moby Dick, had carried Saturn moon rockets to the Cape)—we imagined the truckers to be the heroes of the new highway wilderness. The trucker, like the cowboy, was a figure who summed up the lost freedom of the open spaces, the end of the range, the closing of the frontier. There was no situation more American than the trucker's: always in mid-journey, in-between.

But with the trucker, American drivers shared a sense of the closing road, of limits to time and fuel, of the silting up with circumstance and complication of the whole magic flow of the road.

The crisis saw the boom in citizen's band radios. The name summed up the feeling of drivers: they were just a group of citizens, banded together, speaking their own slang. As a practical matter, CB functioned merely to help drivers avoid the double nickels by warning them when smokies were around. But in another sense it stood for a new sort of camaraderie of the highways, a fellowship of crisis. It was a glorified version of "follow the truckers, they always know the best places to eat."

The truck stop figured as the last survival of the old, independent roadside place. In these old "choke and pukes," the cowboy of the highway gulped down a couple of cups of hundred-mile brew, winked at the waitress, dodged the truck-stop judies and commandoes, and headed back to the rig where his buddy slept in the dog box.

Back on the road he would listen to the constant sound of his radials—"singing waffles," he called them—to country music on the radio or to the airwaves ministry of the Reverend Jimmy Snow, an evangelist who preached to the trucker.

It was a romantic image and one whose fraction of truth was fast fading. By the end of the seventies, computers were put in trucks to second-guess the driver, keeping a log of his speed and fuel consumption, watching over his shoulder to see that he drove neither too fast nor too slow. And before long McDonald's had gone into the truck-stop business.

A human trait that exaggeration has turned into a particularly American one is the preference for building new things to fixing old ones. Politicians call this the "ribbon-cutting syndrome"; it is easier to get a hundred dollars for a new public work than ten dollars to repair an old one. We have always been more willing to build than fix.

In the seventies the neglect came home to roost on the highways. As gasoline revenues fell, maintenance, never very efficiently carried out, declined. Some states had never spent anything on maintaining their Interstates. And some of the Interstates had reached their twenty-year surface-life estimate. Others had deteriorated more rapidly than expected. The transcontinental I-80, for instance, was pitted with holes and the huge cracks truckers call gator backs. Large portions of I-95 needed complete rebuilding. After a bridge across the Mianus River on I-95 in Connecticut collapsed, killing three people, the public realized that there were a quarter of a million "deficient" bridges in their highway system.

The decay had taken place because the federal government did not appropriate a single dollar for maintaining the Interstates until 1976. That had been left to the states; and with federal largess constantly dangling before them, the states found it cheaper and easier to build than to repair.

The completion date of the Interstate system slid further and further into the future. In 1980 an advisory panel to incoming President Ronald Reagan recommended simply stopping—building no more Interstates. To the engineers, this was shocking. What true roadbuilder could ever dare to call the road system finished, or say his job was over? From the original target date of 1972 it had slipped again and again until finally it was 1990. By 1982, the cost of the system had mushroomed from the total of $26 billion planned in 1956 to $275 billion in contemporary dollars. Ony five percent of the system remained to be completed, but those pieces, mostly controversial and complicated urban links, might cost $50 billion. Maintenance of the Interstates annually was estimated at $2 billion; for the rest of the federal system, $5 billion more. All told, federal administrators estimated, the country would have to spend $18 billion a year just to keep the roads from deteriorating further.

Some states, despairing, took advantage of provisions in the law allowing them to "trade in" chunks of Interstate for mass transit money. Whole sections of the ideal map were condemned to fiction.

Weather and wear took their toll of the neglected highways. Nature seemed to be reclaiming the rights of way. Manhattan's West Side Highway, the grand old elevated expressway of the thirties, had received virtually no maintenance for years. One day a truck fell through part of the highway and it was closed to traffic. Soon trees and shrubs grew up in the cracks, and someone painted a huge I-Ching hexagram on the disused pavement.

Even the grandest of superhighways, which had seemed so futuristic when they arrived, were now showing age. The older roads took on the failing, yellow-brown color even the brightest of concrete turns after decades of exposure. What was worse, now that the fascination of the future had worn off, drivers noticed how boring the Interstates were, and how little of the landscape—natural and human—they could see from them.

The energy crisis put an effective period to the great age of roadbuilding. The system was finished. And now, it faced its gravest crisis. Could a closed system satisfy the national need for movement and change? Had the highways been created well enough to institutionalize movement not by simple extension but by repetition?

The whole landscape the highway had created was a continuous denial of centrality, of goal. When Gertrude Stein wrote "There is no there there," she was speaking of Oakland, California, the land at the end of the road. To consider the road having an end, in any sense, was un-American. "The road's what counts. Just look at the road. Don't worry about where it's goin'," says a character in one of Sam Shepard's plays.

The crisis might have been expected to make Americans think about the ends—and the means—of their roads, to question whether it had all been worthwhile. The engineers were sure it had been: they could apply their beloved cost-benefit studies to show that the Interstate system had already paid for itself many times over in lives saved and property values increased and accidents prevented.

Frank Turner even had a specific figure for the savings produced by the Interstates. In 1971, when the cost of the Interstates was calculated at $43 billion, he estimated that by 1977 the system would have saved its users $107 billion. This figure combined all sorts of numbers: a decline by half in the rate of accidents per mile, a saving in lives of one person per year per five miles of Interstate, reduced auto and truck maintenance, and time saved in commercial travel. And if you calculated the savings of time for personal travel at $1.50 an hour, Turner added, the figure rose to $273 billion.

There is no arguing with this sort of calculation, which considers no alternatives and omits the algebra of aesthetics and sociology. But it was not economics that really led us to build the highways in the first place: it was the value we placed on mobility and individual choice.

How can one calculate the values that make a culture decide that it is worthwhile to load the air daily with hundreds of thousands of tons of pollutants, worth the loss of 40,000 human lives a year, or the sacrifice of land equivalent to the state of West Virginia, or—when push comes to shove—the possibility of war to maintain the supply of Middle Eastern petroleum?

It might have made Americans question the gallons of gas burned waiting to pay nickel and dime tolls, made them willing to share cars with their neighbors, made them begin to think of roads as devices for going somewhere, not somewhere to go.

It might have led to a critique of the Highway Trust Fund and the way it operated in a closed, self-regulating loop, a structure of stimulus and response analogous to addiction in a living organism. The more we drove—paying taxes on the gas we burned up and the tires we wore out—the more highways we built. The more highways we had the more we could drive, and so on.

While the Fund had come under attack and efforts had been made to divert some of its revenues to alternative methods of transportation, it took the external shock of the gasoline shortages to disrupt the cycle. Gasoline taxes were levied as fixed amounts per gallon, not as percentage of the price. Thus even as gasoline prices were going up, because Americans were forced by the shortages to drive less, the revenues from the gasoline tax fell.

The crisis could have made them think about all these questions, but it didn't. What it brought out instead was a feeling of betrayal. For as always, all we wanted was to be left alone to drive, with the blue sky above us and open road in front of us, facing an infinite versatility of futures, the palm of fate spread out like a road map on our laps. Was that so much to ask? Hadn't automobility been promised as the fifth freedom?

The deterioration of the highways continued until finally the old political appeals of roadbuilding resurfaced in new clothing, under a chic new term, "infrastructure," that was soon on the lips of every child and old lady. Highways had gotten bad enough that we could talk about "rebuilding" them rather than "repairing" them. So the gas tax was raised from four to nine cents a gallon. The bond traders cranked up the tax-exempts. The big orange "Your tax dollars at work" signs, with the names of the relevant politicians, began to reappear, and the delays and detours of construction replaced those of potholes and gasoline lines. Here was a chance to build the highways over again, at a cost, by one industry estimate, of $325 billion over thirty-two years. Confronted with the end of the road, Americans did what they had always done: they started over.

The clearest indication of just how much the gasoline crisis upset the country was the speed with which it forgot it, the force with which it repressed the memory. In 1980, Ronald Reagan ran on a Republican platform that called for the abolition of the fifty-five-mile-an-hour speed limit. Gasoline consumption had been cut, traffic fatalities had begun to taper off, but Americans quickly reverted to the old ways. Detroit brought out a whole new line of "muscle cars," more fuel-chary than their predecessors but boasting of their zero-to-sixty times and their performance in the passing lane. Recreation vehicle sales crept back up and so did the statistics on miles traveled.

Many of the watchers were still trapped in traffic, miles away, when the shuttle lifted off, and they were still there when traffic began to flow, ever so slowly, the other way. By then, the shuttle had passed completely around the earth and back almost overhead, and the astronauts could watch the traffic creeping along.

From space, the traffic and the pattern of the roads were the most prominent visible feature of civilization, the circulatory system of America, and its versions in Europe, in Japan, in showcase freeways in the Third World.

The astronaut knows what he is seeing, but for a visitor approaching the planet from even further away, say from another star system, the shapes would be more striking, more puzzling.

The alien eye seeks patterns and connections: Earthlings imagined a Mars lined with canals. It would be quite understandable if some arriving alien, viewing first the roads, and then, as he came closer, picking out the vehicles on them, might conclude that the planet's dominant form of life was mobile glass and steel, and its supporting environment the streams and rivers of asphalt and concrete. But he would be hard put to arrive at a satisfactory hypothesis of the nature of its nutritional process or the organizing principle of its lines and patterns. And least of all could he imagine, in his wildest speculations, what might be the goal and purpose of this coursing, relentless flow.

Acknowledgments

Of the many people who helped in the research for this book, I would particularly like to thank Mary Jo Burke at the Department of Transportation Library, William Bridges of Kentucky Fried Chicken, and John Margolies. Francis Turner, former federal highway administrator and "Mr. Interstate," was generous with his time and recollections. Reed Estabrook discussed his photographic journey along I-80. Roger White of the National Museum of American History, Transportation Department, shared his files and insights. Frank Simonini of the New Jersey Department of Transportation let me look at the site of construction on I-78. Geoffrey Orton and his colleagues at the New York Department of Transportation allowed me to see the computers used in highway design.

For suggestions and support in the research and associated projects, my thanks go to Domenico Annese, Joshua Baskin, John Boddie, Kim Brown, Jennifer Crandall, John Crawford, Elizabeth Fore,

Alan Hess, Jerry Lazar, Anita Leclerc, Ronni Lundy, John Mariani, Doug Michels, Tom Passavant, Ben Schiff, Neil Selkirk, Maggie Simmons, Richard Story, David Walker, Michaela Williams, and William Wilson. Melanie Jackson, my agent, and three able editors, Morgan Entrekin, John Herman, and Jane Isay, guided the project from initial idea to finished book.

My greatest debt is to my wife Joëlle for her patience and support and for her sharp suggestions and readings.

Notes on Sources

Here, selectively listed and described, are some background sources for each chapter. Specific references made in the text are not repeated.

For the basic history of the automobile and highway, the two best academic introductions are John Rae's *The Road and the Car in American Life* and James Flink's *The Car Culture*. For travel on the highway, the volumes are few: Warren Belasco's *Americans on the Road* and John Jakles's *The Tourist*. *The Highway and the Landscape* is a series of essays on the development of American road design, edited by W. Brewster Snow. Christopher Tunnard and Boris Pushkarev's *Man-Made America* is a fundamental study of the built landscape. Lewis Mumford is indispensable background, along with Vincent Scully's *American Architecture and Urbanism* and William H. Whyte's *The Last Landscape*. Leo Marx's classic *The Machine in the Garden* and Roderick Nash's *Wilderness and the American Mind* are basic interpretations of American attitudes toward what we today call "development." For the American obsession with mobility, see George Pierson's *The Moving American*.

Chapter 1. The Road into the Wilderness
The grid system is discussed many places in J.B. Jackson's books, and in John R. Stilgoe's *Common Landscape of America*. Debates over internal improvements are treated in most common histories of the period; see also Robert Remini's biography of Andrew Jackson and Henry Adams's biography of Albert Gallatin. For John C. Calhoun, the booster of roads and railroads before he became known as a secessionist, see Richard Hofstadter's essay, "The Marx of the Master Class," in *The American Political Tradition*.

Chapter 2. Boosters and Nostalgists
The story of the Lincoln Highway as told by its founders is found in *The Lincoln Highway: The Story of a Crusade That Made Transportation History*. See also Joe McCarthy's article in *American Heritage*, June 1974.

Chapter 3. The Kingfish's Highways, or Everyman's Locomotive
Huey Long's roadbuilding program is documented in T. Harry Williams's definitive biography. Long's own justifications for it are laid out in his "autobiography" and tract, *Every Man a King*. For the impact of the road in the South, see William Faulkner, especially *Sartoris*, and Flannery O'Connor, *Wise Blood* and *Collected Stories*. Blaine Brownell gives the academic perspective on the development of highways in southern cities—boosted by the Chamber of Commerce and less formal businessmen's organizations—in *The Urban Ethos in the South, 1920–1930*. The Chevrolet ad is reproduced in Jane and Michael Stern's *Auto Ads*.

Chapter 4. Parkway to Freeway
Fitzgerald's Dr. Eckleburg is discussed by Karal Ann Marling in *The Colossus of Roads*. For the development of the parkway see Carl Condit, *American Building Art*, and *Man-Made America* by Tunnard and Pushkarev. Robert Caro's *The Power Broker* tells the story of Robert Moses's constructions. Benton MacKaye's articles on the Townless Highway appeared in *The New Republic*, March 12, 1930, and *Harper's*, August 1931.

Chapter 5. Ike's Autobahns
For the Autobahns, there are Hans Lorenz's volumes in German, Karl Larmer's *Autobahnbau in Deutschland*, and for a first-hand description of the prewar system, F.A. Gutheim's article in *American Magazine of Art*, April 1936. Several first-hand American accounts of Autobahn construction are in the DOT Library. The best, if often turgid and complicated, account of the passage of the Interstate legislation is Mark Rose, *Interstate: Express Highway Politics*. For the Clay Commission report, see *The Age of Asphalt*, a volume of documents edited by Richard O. Davis. Also of interest is Mark Foster, *From Street Car to Superhighway*. I have also relied on interviews with Francis Turner and others for this section.

Chapter 6. The Superhighway in the City
The best look at the Los Angeles freeways is David Brodsly's *LA Freeway*, although Randolph Collier and other builders are given short shrift. Reyner Banham and Charles Moore are two excellent writers on the freeways as architecture. Banham's *Megastructure: Urban Futures of the Recent Past* looks at such projects as the Lower Manhattan Expressway. The liberal reaction to the highway program is exemplified by Helen Leavitt's *Superhighway, Super Hoax*, Ben Kelley's *The Pavers and the Paved*, and A.Q. Mowbray's *Road to Ruin*. Some fallacies of mass transit are discussed in J. Allen Whitt, *Urban Elites and Mass Transportation*.

Chapter 7. Streamline to Assembly Line
Norman Bel Geddes's books are documents of their time; some of the best interpretations of the period are Donald Bush's *The Streamlined Decade* and Jeffrey Meikle's fine *Twentieth Century Limited*, with its discussion of Miller McClintock's traffic flow theories.

Chapter 8. *Mies en scène*: The Shape of the Superhighway

Chapter 9. The Triumph of the Engineers
The DOT Bicentennial volume offers something like an official line on the history of highway engineering. See also Carl Condit's *Amer-*

ican Building Art, John Robinson's *Highways and Our Environment*, and the ever-useful *Man-Made America*. John Kouwenhoven's *Made in America* is a classic on what is American about American design. Horatio Greenough's writings are most easily available in *Form and Function: Remarks on Art, Design, and Architecture*, edited by Harold A. Small. For a good discussion of horizontality in Frank Lloyd Wright, see Peter Blake's *The Master Builders*. Lawrence Halprin's *Freeways* provides a designer's own perspective. *The View from the Road* by Donald Appleyard et al. attempts to devise a system for coding the views from highways and proposes a way to locate them for improved aesthetics, while *The Freeway in the City*, prepared by the urban advisors to the Federal Highway Administration, is a sixties consensus look at how to build urban highways. The New York, New Jersey, and Texas Departments of Transportation and the Intergraph Corporation were also helpful in preparing these chapters.

Chapter 10. The Great Striptease

Chapter 11. Signs
Recent years have seen a spate of books on roadside architecture: *California Crazy* by Jim Heimann and Rip Georges, *The End of the Road* by John Margolies, and *The Well-Built Elephant and Other Roadside Attractions* by J.J.C. Andrews. John Baeder's wonderful *Diners* and his *Gas, Food, and Lodging* are made up mostly of postcards from his own collection. Chester Liebs's *Roadside Architecture from Main Street to Miracle Mile* is a solid job. Daniel Vieyra's *Fill 'er Up* discusses gas stations. Sally Henderson's *Billboard Art* is more entertaining than historical.

Chapter 12. Bypass
For the story of Colonel Sanders see his autobiography, *Life As I Have Known It*, and John Ed Pearce, *The Colonel*. I have also relied on a number of interviews and the files of the company, including *The Bucket*, the franchisee's magazine.

Chapter 13. Franchised America
Many of the major highway franchisers are notoriously without
sense of history. Howard Johnson's is an example; much of the
information on its early years comes from *Fortune*, September 1940,
and from Warren Belasco's article in *Journal of American Culture*,
Fall 1979, called "Toward a Culinary Common Denominator:
Origins of Howard Johnson's." The standard version of the
McDonald's story is Ray Kroc's *Grinding It Out*. Max Boas and
Steve Chain give another, "unauthorized" but rarely critical, view
in *Big Mac*. Alan Hess is the authority on the early architecture of
McDonald's, summarized in "Golden Architecture," *LAICA Journal*, Spring 1983. Materials from the McDonald's public relations
department were also helpful. The Holiday Inn story is told in
Time, June 12, 1972, and in the autobiography of Kemmons Wilson's partner, Wallace Johnson, *Work Is My Play*. For an overview
of the franchises, there is Stan Luxenberg's *Roadside Empires: How
the Chains Franchised America*. *White Towers* by Steven Izenour and
Paul Hirshorn is a wonderful photographic survey of the architecture of one unfranchised chain.

Chapter 14. The New Highway Landscape
Peter O. Muller's *Contemporary Suburban America* is an overview of
some of the changes wrought by the highways.

Chapter 15. Why A Duck?: The Redemption of Roadside Architecture
Robert Venturi's writings are well known; James Wines's statement
on the duck is found in "The Case for the Big Duck," *Architectural
Forum*, April 1972. See also "The Big Duck" by Howard Mansfield
in *Sites* XII, 1984. For Wines's McDonald's, see the Chrysler Museum's catalog by Thomas Sokolowski.

Chapter 16. Sentimental Journeys: The Myths of Route 66
For Route 66 itself, there are many articles. See Kerouac's *On the
Road*, and for Ken Kesey Tom Wolfe's *The Electric Kool-Aid Acid
Test*. Alfred Appel's *Nabokov's Dark Cinema* and *Signs of Life* are both

quirky and entertaining. Roosevelt's advice to take to the road is quoted in William Stott, *Documentary Expression and Thirties America*, which also includes an excellent discussion of the New Deal's documentary programs.

Chapter 17. Escape Routes

Chapter 18. The Mobile Eye: The Camera on the Road

Chapter 19. Ends of the Road
The figures cited for highway reconstruction are from the Department of Transportation's annual reports and The Road Information Project, an industry organization and lobby.

Bibliography

BOOKS

Adams, Robert. *The New West*. Colorado Associated University Press, 1974.

Agee, James. *A Death in the Family*. Bantam, 1969.

Andrews, J.J.C. *The Well-Built Elephant and Other Roadside Attractions*. Congdon & Weed, 1984.

Ant Farm [Chip Lord, Doug Michels, and others]. *AutoAmerica: A Trip Down U.S. Highways from World War II to the Future*. Dutton, 1976.

Appel, Alfred. *Nabokov's Dark Cinema*. Oxford University Press, 1974.

———. *Signs of Life*. Knopf, 1983.

Appleyard, Donald, Kevin Lynch, and John R. Myer. *The View from the Road*. MIT Press, 1964.

Baeder, John. *Diners*. Abrams, 1979.

———. *Gas, Food, and Lodging*. Abbeville, 1982.

Baker, Geoffrey and Bruno Funaro. *Motels*. Reinhold, 1954.

Baltz, Lewis. *Park City*. Aperture, 1980.

Banham, Reyner. *Los Angeles: The Ecology of Four Environments*. Penguin, 1971.

———. *The Ecology of the Well-Tempered Environment*. The University of Chicago Press, 1969.

———. *Theory and Design in the First Machine Age*. Praeger, 1967.

———. *Megastructure: Urban Futures of the Recent Past*. Harper & Row, 1976.

Belasco, Warren. *Americans on the Road: From Autocamp to Motel, 1910–1945*. MIT Press, 1979.

Bel Geddes, Norman. *Horizons*. Random House, 1932.

———. *Magic Motorways*. Random House, 1940.

———. *Miracle in the Evening*. Ed. William Kelley, Doubleday, 1960.

Billington, David. *The Tower and Bridge*. Basic, 1983.

Blake, Peter. *God's Own Junkyard*. Holt, Rinehart and Winston, 1964.

———. *The Master Builders*. Norton, 1976.

Boas, Max and Steve Chain. *Big Mac*. Dutton, 1976.

Boorstin, Daniel. *The Americans: The Democratic Experience*. Vintage, 1974.

———. *The Image: A Guide to Pseudo-Events in America*. Harper, 1961.

Bruce-Briggs, B. *The War against the Automobile*. Dutton, 1977.

Brodsly, David. *LA Freeway: An Appreciative Essay*. University of California Press, 1982.

Brown, Dee. *Hear that Lonesome Whistle Blow: Railroads in the West*. Holt, Rinehart and Winston, 1979.

Brownell, Blaine A. *The Urban Ethos in the South, 1920–1930*. Louisiana State University Press, 1975.

Bush, Donald J. *The Streamlined Decade*. Braziller, 1975.

Cain, James. M. *Mildred Pierce*. Knopf, 1941.

———. *The Postman Always Rings Twice*. Vintage, 1978 (1934).

California Division of Highways. *How Los Angeles Was Unified by Freeways*. State of California, 1966.

Caro, Robert. *The Power Broker: Robert Moses and the Fall of New York*. Random House, 1975.

Chevallier, Raymond. *Roman Roads*. University of California Press, 1976.

Clay, Grady. *Close-Up: How to Read the American City*. Praeger, 1973.

Collier, Randolph. *"The Legislature Takes a Look at California Highway Needs" and Other Addresses*. California State Senate, 1947.

Condit, Carl. American Building Art. Oxford University Press, 1961.

Davis, Richard O., Ed. *The Age of Asphalt: The Automobile, the Freeway, and the Condition of Metropolitan America.* Lippincott, 1975.

Dettelbach, Cynthia G. *In the Driver's Seat: The Automobile in American Literature and Popular Culture.* Greenwood Press, 1976.

Didion, Joan. *Play It As It Lays.* Bantam, 1970.

————. *The White Album.* Simon & Schuster, 1979.

Donavan, Frank. *Wheels for a Nation.* Thomas Y. Crowell, 1965.

Dreiser, Theodore. *A Hoosier Holiday.* Lane, 1916.

Drexler, Arthur. *Engineering.* The Museum of Modern Art, 1964.

Emerson, Ralph Waldo. *Selected Essays, Lectures, and Poems.* Washington Square Press, 1965.

Evans, Walker. *American Photographs.* Museum of Modern Art, 1938.

Fein, Albert. *Frederick Law Olmsted and the American Environmental Tradition.* Braziller, 1972.

Fitzgerald, F. Scott. *The Great Gatsby.* Scribner's, 1925.

Flink, James. J. *America Adopts the Automobile, 1895–1910.* MIT Press, 1970.

————. *The Car Culture.* MIT Press, 1975.

Foster, Mark. *From Street Car to Superhighway.* Temple University Press, 1981.

Frank, Robert. *The Americans.* Grove, 1959.

Friedlander, Lee. *Self-Portrait.* Haywire, 1970.

Gallatin, Albert. "Report on Roads and Canals," in *Writings.* Ed. Henry Adams, 1879.

Giedion, Siegfried. *Space, Time, and Architecture.* Harvard University Press, 1940.

————. *Mechanization Takes Command.* Oxford University Press, 1948.

Goldfield, David and Blaine Brownell. *Urban America: From Downtown to No Town.* Houghton Mifflin, 1979.

Gordon, Jan and Cora J. Gordon. *On Wandering Wheels: Through Roadside Camps from Maine to Georgia in an Old Sedan Car.* Dodd, Mead, 1928.

Greenough, Horatio. *Form and Function: Remarks on Art, Design, and Architecture.* Ed. Harold A. Small. University of California Press, 1947.

Gruen, Victor and Larry Smith. *Shopping Towns USA: The Planning of Shopping Centers.* Reinhold, 1960.

Gutman, Richard and Elliott Kaufman with David Slovic. *American Diner*. Harper and Row, 1979.

Halprin, Lawrence. *Freeways*. Reinhold, 1966.

Harbison, Robert. *Eccentric Spaces*. Knopf, 1977.

Heimann, Jim and Rip Georges. *California Crazy: Roadside Vernacular American Architecture*. Chronicle, 1980.

Henderson, Sally and Robert Landau. *Billboard Art*. Chronicle, 1981.

Hitchcock, Henry-Russell. *The Architecture of H. H. Richardson and His Times*. MIT Press, 1961.

Hofstadter, Richard. *The American Political Tradition and the Men Who Made It*. Knopf, 1948.

Izenour, Steven and Paul Hirshorn. *White Towers*. MIT Press, 1979.

Jackson, John Brinckerhoff. *Discovering the Vernacular Landscape*. Yale University Press, 1984.

———. *Landscapes: Selected Writings of J.B. Jackson*. Ed. Erwin Zube. University of Massachusetts Press, 1970.

Jackson, William Turrentine. *Wagon Roads West*. Yale University Press, 1965.

Jacobs, Jane. *The Death and Life of Great American Cities*. Random House, 1961.

———. *The Economy of Cities*. Random House, 1969.

Jakles, John. *The Tourist: Travel in Twentieth Century North America*. University of Nebraska Press, 1985.

James, Henry. *The American Scene*. Harper, 1907.

Jefferson, Thomas. *The Portable Thomas Jefferson*. Ed. Merrill D. Peterson. Viking, 1975.

Jerome, John. *The Death of the Automobile*. Norton, 1972.

Johnson, Wallace E. *Work Is My Play*. Hawthorn, 1973.

Keats, John. *The Insolent Chariots*. J. B. Lippincott Co., 1958.

Kelley, Ben. *The Pavers and the Paved*. D. W. Brown, 1971.

Kerouac, Jack. *On the Road*. Signet, 1957.

Keyton, Clara. *Tourist Camp Pioneering Experiences*. Adams Press (Chicago), 1960.

Kouwenhoven, John. *The Beer Can by the Highway*. Doubleday, 1960.

———. *Made in America* (republished as *The Arts in American Civilization*). Doubleday, 1948.

Kowinski, William Severini. *The Malling of America: An Inside Look at the Great Consumer Paradise.* Morrow, 1985.

Kroc, Ray. *Grinding It Out.* Contemporary, 1977.

Kurtz, Stephen A. *Wasteland: Building the American Dream.* Praeger, 1973.

Laas, William, Ed. *Freedom of the American Road.* Ford Motor Company, 1956.

Labatut, Jean and Wheaton J. Lane, Eds. *Highways in Our National Life.* Princeton University Press, 1950.

Lange, Dorothea and Paul Taylor. *American Exodus.* Reynal & Hitchcock, 1939.

Larmer, Karl. *Autobahnbau in Deutschland, 1933–1945.*

Least Heat Moon, William [William Trogdon]. *Blue Highways: A Journey into America.* Houghton Mifflin, 1982.

Leavitt, Helen. *Superhighways Superhoax.* Doubleday, 1970.

Leonhardt, Fritz. *Bridges: Aesthetics and Design.* MIT Press, 1982.

Lewis, David L. and Laurence Goldstine, Eds. *The Automobile and American Culture.* University of Michigan Press, 1980.

Lewis, Sinclair. *Babbitt.* Harcourt, Brace, 1922.

———. *Main Street.* Harcourt, Brace, 1920.

———. *Free Air.* Grosset & Dunlap, 1919.

———. *The Man Who Knew Coolidge.* Harcourt, Brace, 1941.

Liebs, Chester. *Roadside Architecture: From Main Street to Miracle Mile.* New York Graphic, 1986.

Lincoln Highway Association. *The Lincoln Highway: The Story of a Crusade That Made Transportation History.* Dodd, Mead, 1935.

Lingeman, Richard. *Small Town America: A Narrative History, 1620–The Present.* Putnam's, 1980.

Long, Huey P. *Every Man a King.* (New Orleans) 1933.

Lorenz, Hans and F. A. Finger, Eds. *Trassierungsgrundlagen der Reichsautobahnen.* Volk und Reich (Berlin), 1943.

———. *Trassierung und Gestaltung von Strassen und Autobahnen.* Volk und Reich, 1943.

Luxenberg, Stan. *Roadside Empires: How the Chains Franchised America.* Viking, 1985.

Lynch, Kevin. *The Image of the City.* MIT Press, 1960.

MacCannell, Dean. *The Tourist: A New Theory of the Leisure Class.* Schocken, 1976.

Mallory, Keith and Arvid Ottan. *The Architecture of War.* Pantheon, 1973.

Malone, Dumas. *Thomas Jefferson and His Times*. Little, Brown, 1981.

Margolies, John. *The End of the Road: Vanishing Highway Architecture in America*. Penguin, 1981.

Marling, Karal Ann. *The Colossus of Roads: Myth and Symbol along the American Highway*. University of Minnesota Press, 1984.

Marx, Leo. *The Machine in the Garden*. Oxford Univesity Press, 1964.

Mason, Philip P. *A History of American Roads*. Rand McNally, 1967.

————. *The League of American Wheelmen and the Good Roads Movement, 1880–1905*. University of Michigan Press, 1958.

McLuhan, Marshall. *Understanding Media*. McGraw-Hill, 1964.

Meikle, Jeffrey. *Twentieth Century Limited: Industrial Design in America, 1925–1939*. Temple University Press, 1979.

Miller, Henry. *The Air-Conditioned Nightmare*. New Directions, 1945.

Moore, Charles, Peter Baker, and Regula Campbell. *Los Angeles: The City Observed: A Guide to Its Architecture and Landscapes*. Vintage, 1984.

Mowbray, A. Q. *Road to Ruin*. J. B. Lippincott, 1969.

Muller, Peter O. *Contemporary Suburban America*. Prentice-Hall, 1981.

Mumford, Lewis, *The City in History*. Harcourt, 1961.

————. *The Highway and the City*. Harcourt, 1956.

————. *Technics and Civilization*. Harcourt, 1934.

Nabokov, Vladimir. *Lolita*. Olympia Press, 1955.

Nader, Ralph. *Unsafe at Any Speed*. Grossman, 1965.

Nairn, Ian. *The American Landscape*. Random House, 1965.

Nash, Roderick. *Wilderness and the American Mind*. Yale University Press, 1979.

Nevins, Alan. *Ford*. Scribner's, 1954.

Norris, Frank. *The Octopus: A Story of California*. Houghton Mifflin, 1958 (1901).

O'Connor, Flannery. *Wise Blood*. Signet, 1953.

Onosko, Tim. *Wasn't the Future Wonderful?* Dutton, 1979.

Parkman, Francis. *The Oregon Trail*. Penguin, 1982 (1849).

Pearce, John, Ed. *The Colonel*. Doubleday, 1982.

Perkins, Pamel Gruninger. *Autoscape: The Automobile in the American Landscape* [exhibition catalog]. Whitney Museum of Art, Fairfield County, 1984.

Pettifer, Julian and Nigel Turner. *Automania: Man and the Motor Car*. Little, Brown, 1985.

Pierson, George. *The Moving American*. Knopf, 1973.

Pomeroy, Earl. *In Search of the Golden West: The Tourist in Western America*. Knopf, 1957.

Rae, John. *The Road and the Car in American Life*. MIT Press, 1971.

Rand, Christopher. *Los Angeles: The Ultimate City*. Oxford University Press, 1967.

Rittenhouse, Jack. *Guide to Highway 66*. Self-published, Los Angeles, 1946.

Robinson, John. *Highways and Our Environment*. McGraw-Hill, 1971.

Rose, Mark. *Interstate: Express Highway Politics, 1941–1956*. The Regents Press of Kansas (Lawrence, Kansas), 1979.

Rothschild, Emma. *Paradise Lost: The Decline of the Auto-Industrial Age*. Random House, 1973.

Ruscha, Ed. *Every Building on the Sunset Strip*. 1966.

———. *Thirty-four Parking Lots*. 1967.

Sanders, Harland. *Life As I Have Known It Has Been Finger-Licking Good*. Creation House (Carol Stream, Illinois), 1974.

Schickel, Richard. *The Disney Version*. Simon & Schuster, 1968.

Scully, Vincent. *American Architecture and Urbanism*. Praeger, 1969.

Shank, William H. *Vanderbilt's Folly: A History of the Pennsylvania Turnpike*. American Canal and Transportation Center, 1973.

Shore, Stephen. *Uncommon Places*. Aperture, 1982.

Sindler, Allan. *Huey Long's Louisiana*. Johns Hopkins University Press, 1956.

Sky, Allison and Michelle Stone. *Unbuilt America*. McGraw-Hill, 1976.

Sloan, Alfred. *My Years With General Motors*. Doubleday, 1964.

Smith, Henry Nash. *Virgin Land: The American West as Symbol and Myth*. Harvard University Press, 1950.

Snow, W. Brewster, Ed. *The Highway and the Landscape*. Rutgers University Press, 1959.

Sokolowski, Thomas W. *SITE on McDonald's: The American Landscape*. The Chrysler Museum, 1984.

Speer, Albert. *Spandau Diaries*. Macmillan, 1976.

Stein, Gertrude. *Lectures in America*. Random House, 1935.

Steinbeck, John. *The Grapes of Wrath*. Bantam, 1946 (1939).

———. *Travels with Charley: In Search of America*. Penguin, 1980.

Stern, Jane and Michael Stern. *Trucker*. McGraw Hill, 1975.

———. *Amazing America*. Random House, 1978.

———. *Auto Ads*. Random House, 1978.

Stewart, George. *US 40: Cross Section of the US*. Houghton Mifflin Co., 1953.

Stilgoe, John R. *Common Landscape of America: 1580 to 1845*. Yale University Press, 1982.

———. *Metropolitan Corridor*. Yale University Press, 1984.

Stott, William. *Documentary Expression in Thirties America*. Oxford University Press, 1973.

Susman, Warren I. *Culture as History: The Transformation of American Society in the Twentieth Century*. Pantheon, 1984.

Szarkowski, John. *Walker Evans*. Museum of Modern Art, 1971.

de Tocqueville, Alexis. *Democracy in America*. Trans. Henry Reeve. Knopf, 1945.

Trachtenberg, Alan. *Brooklyn Bridge, Fact and Symbol*. University of Chicago Press, 1979.

Tunnard, Christopher and Boris Pushkarev. *Man-made America: Chaos or Control?* Yale University Press, 1963.

Urban Advisors to the Federal Highway Administration (Michael Rapuano et al.) *The Freeway in the City: Principles of Planning and Design*. Government Printing Office, 1968.

U.S. Department of Transportation. *America's Highways 1776–1976*. Government Printing Office, 1976.

———. *The Status of the Nation's Highways: Conditions and Performance*. Government Printing Office, 1985.

U.S. Federal Highway Administration. *Standard Plans for Highway Bridges*. Government Printing Office, 1968.

Venturi, Robert. *Complexity and Contradiction in Architecture*. Museum of Modern Art, 1966.

———. *The Highway* [exhibition catalog]. Institute of Contemporary Art, University of Pennsylvania, 1970.

Venturi, Robert, Denise Scott Brown, and Steven Izenour. *Learning from Las Vegas*. MIT Press, 1972.

Vieyra, Daniel. *Fill 'er Up: An Architectural History of America's Gas Stations*. Macmillan, 1979.

Warner, Sam Bass. *The Urban Wilderness: A History of the American City*. Harper & Row, 1972.

Warren, Robert Penn. *All the King's Men.* Harcourt, Brace, 1946.

Whitman, Walt. *Leaves of Grass and Selected Prose.* Rinehart, 1949.

Whitt, J. Allen. *Urban Elites and Mass Transportation: The Dialectics of Power.* Princeton University Press, 1982.

Whyte, William. *The Last Landscape.* Doubleday, 1968.

Wik, Reynold M. *Henry Ford and Grass Roots America.* University of Michigan, 1972.

Williams, T. Harry. *Huey Long.* Knopf, 1964.

Wilson, Kemmons. *The Holiday Inn Story.* Holiday Press, 1973.

Wolfe, Tom. *The Electric Kool-Aid Acid Test.* Farrar, Straus and Giroux, 1968.

Works Progress Administration Guides.

Wright, Frank Lloyd. *The Disappearing City.* W. F. Payson, 1932.

———. *An Autobiography.* Duell Sloan and Pearce, 1943.

Zelinsky, Wilbur. *The Cultural Geography of the U.S.* Prentice-Hall, 1973.

Zube, Ervin H. and Margaret J. Zube, Eds. *Changing Rural Landscapes.* University of Massachusetts Press, 1977.

———. *Fritz Todt: Mensch, Ingenieur, Nationalsozialist.* Verlag Gerhard Stallic-Oldenburg, 1943.

PERIODICALS

[Agee, James] "The Great American Roadside." *Fortune*, September 1934.

Alloway, Lawrence. "Hiway Culture." *Arts*, February 1967.

Banham, Reyner. "The Missing Motel." *Landscape* 15, 1965.

Belasco, Warren. "Toward a Culinary Common Denominator: Origins of Howard Johnson's." *Journal of American Culture*, Fall 1979.

Bourke-White, Margaret. "Road Signs." *Life*, June 27, 1938.

Cameron, Juan. "How the Interstate Changed the Face of the Nation." *Fortune*, July 1971.

Ciotti, Paul. "Who Needs Metrorail?" *California*, March 1984.

Crewdson, John M. "The Interstate's Shadow Is Overtaking Route 66." *The New York Times*, July 7, 1981.

Crozier, Emmet. "How Spectacle Came Into Being." *The New York Herald-Tribune*, July 18, 1939.

Dunne, John Gregory. "Eureka!" *New West*, January 1, 1979.

Duffus, R. L. "The Highways of the Future." *The New York Times Book Review*, March 14, 1940.

Evans, Walker. Interview. *New Republic*, November 1976.

Fortune. "The Howard Johnson Restaurants." September 1940.

Fortune. "The US Highway System." June 1941.

Galbraith, John Kenneth. "To My New Friends in the Affluent Society—Greetings!" *Life*, March 27, 1970.

Gannett, Lewis. "Magic Motorways" (review). *The New York Herald-Tribune*, March 12, 1940.

Geist, William E. "Drive-in Movies." *The New York Times*, June 7, 1983.

Greenland, Drew. "Route 66." *Life*, June 1983.

Gutheim, F.A. "German Highway Design: The Reichsautobahn." *American Magazine of Art*, April 1936.

Hammonds, Keith. "Drive-ins." *The New York Times*, May 30, 1982.

Herbers, John. "Urban Centers' Population Drift Creating a Countryside Harvest." *The New York Times*, March 23, 1980.

Hess, Alan. "Golden Architecture." *LAICA Journal*, Spring 1983.

Holsendolph, Ernest. "Outdoor Advertisers Read the Small Print." *The New York Times*, October 19, 1980.

Hoover, J. Edgar. "Camps of Crime." *American Magazine*, February 1940.

Howarth, William. "The Okies: Beyond the Dust Bowl." *National Geographic*, September 1984.

Hubbell, Sue. "On the Interstate: A City of the Mind." *Time*, June 3, 1985.

Jackson, J. B. "Urban Circumstances." *Design Quarterly* 128, 1985.

Jackson, J. B. "Garages." *Landscape*, Winter 1976.

Kaempffert, Waldemar. "Magic Carpet in Futurama." *The New York Times*, September 10, 1939.

Kwitny, Jonathan. "The Great Transportation Conspiracy." *Harper's*, February 1981.

McCarthy, Joe. "The Lincoln Highway." *American Heritage*, June 1974.

McFadden, Robert D. "I-95 Bridge Closed." *The New York Times*, September 1, 1984.

MacKaye, Benton. "The Townless Highway." *The New Republic*, March 12, 1930.

MacKaye, Benton and Lewis Mumford. "Townless Highways for the Motorist." *Harper's*, August 1931.

Mansfield, Howard. "The Big Duck." *Sites* XII, 1984.

Moses, Robert. "New Highways for a Better New York," *The New York Times Magazine*, November 11, 1945.

New West. "California Highways." September 8, 1980.

The New York Times. "Other Projects Get $5 Billion in Interstate Road 'Trades.'" December 3, 1978.

————. "General Motors Host to Employees." May 20, 1939.

————. "Fair Visitors 'Fly' Over U.S. Of 1960." April 19, 1939.

————. "Auto Trip to Coast in a Day Predicted." May 16, 1939.

————. "Super-Highways." January 28, 1940.

Peterson, Iver. "The End of the Road." *Rolling Stone*, November 22, 1984.

Pew, Thomas W. "Route 66." *American Heritage*, August 1977.

Phelps, Dawson A. "The Natchez Trace: Indian Trail to Parkway." *Tennessee Historical Quarterly*, September 1962.

Reader's Digest. "Our Great Big Highway Bungle." July 1960.

Robbins, L.H. "America Hobnobs at the Tourist Camp." *The New York Times Magazine*, August 12, 1934.

Roberts, Steven. "Ode to a Freeway." *The New York Times Magazine*, April 15, 1973.

Time, "Rapid Rise of the Host with the Most." June 12, 1972.

USA Today. "Red-Eyed Pete's Demise Opens Way for Progress." November 15, 1983.

Whitworth, William. "Kentucky-Fried." *The New Yorker*, February 14, 1970.

Wines, James. "The Case for the Big Duck." *Architectural Forum*, April 1972.

Index